Algorithms for Sample Preparation with Microfluidic Lab-on-Chip

RIVER PUBLISHERS SERIES IN BIOMEDICAL ENGINEERING

Indexing: All books published in this series are submitted to the Web of Science Book Citation Index (BkCI), to SCOPUS, to CrossRef and to Google Scholar for evaluation and indexing.

The "River Publishers Series in Biomedical Engineering" is a series of comprehensive academic and professional books which focus on the engineering and mathematics in medicine and biology. The series presents innovative experimental science and technological development in the biomedical field as well as clinical application of new developments.

Books published in the series include research monographs, edited volumes, handbooks and textbooks. The books provide professionals, researchers, educators, and advanced students in the field with an invaluable insight into the latest research and developments.

Topics covered in the series include, but are by no means restricted to the following:

- Biomedical engineering
- Biomedical physics and applied biophysics
- Bio-informatics
- Bio-metrics
- Bio-signals
- Medical Imaging

For a list of other books in this series, visit www.riverpublishers.com

Algorithms for Sample Preparation with Microfluidic Lab-on-Chip

Sukanta Bhattacharjee
Indian Statistical Institute
India

Bhargab B. Bhattacharya
Indian Statistical Institute
India

Krishnendu Chakrabarty
Duke University
USA

LONDON AND NEW YORK

Published 2018 by River Publishers
River Publishers
Alsbjergvej 10, 9260 Gistrup, Denmark
www.riverpublishers.com

Distributed exclusively by Routledge
4 Park Square, Milton Park, Abingdon, Oxon OX14 4RN
605 Third Avenue, New York, NY 10017, USA

Algorithms for Sample Preparation with Microfluidic Lab-on-Chip / by Sukanta Bhattacharjee, Bhargab B. Bhattacharya, Krishnendu Chakrabarty.

Routledge is an imprint of the Taylor & Francis Group, an informa business

ISBN 978-87-7022-055-2 (print)

Contents

Preface

Microfluidic technologies are one of the major driving force towards the miniaturization of laboratory-based biochemical protocols. Rapid sample processing, usage of nano- or pico-liter range sample volume, and the capability of precise control of fluids in an assay have made microfluidic chips an attractive choice to replace traditional bench-top diagnostics procedures with a miniaturized total analysis systems or lab-on-a-chip (LoC). Especially, in the era of personalized medicine, microfluidic-based LoC devices will provide a ubiquitous platform for rapid diagnostics and precision care. Biochips have already made a profound impact on various application domains such as clinical diagnostics, DNA analysis, genetic engineering, and drug discovery, among others.

In almost all bio-protocols, sample preparation plays an important role which includes dilution and mixing of several fluids satisfying certain volumetric ratios. However, designing algorithms that minimizes reactant-cost and sample-preparation time suited for microfluidic chips poses a great challenge from the perspective of protocol mapping, scheduling, and physical design. This monograph undertakes to bridge the widening gap between biologists and engineers by introducing, from the fundamentals, several state-of-the-art computer-aided-design algorithms for sample preparation with digital and flow-based microfluidic biochips. By this, the monograph summarizes the results of several years of research efforts at Indian Statistical Institute in collaboration with Prof. Tsung-Yi Ho from the National Tsing Hua University, Taiwan, Prof. Juinn-Dar Huang from the National Chiao Tung University, Taiwan, Prof. Robert Wille from the Johannes Kepler University Linz, Austria, and Prof. Sudip Roy from the Indian Institute of Technology Roorkee, India. We would like to take the opportunity to acknowledge them for their valuable suggestions. Besides that, we are thankful to the coauthors of the research papers which constituted the basis of this book, including Prof. Ansuman Banerjee, Sudip Poddar, and Yi-Ling Chen. Finally, we thank River Publisher and, in specific, Mark de Jongh for making this monograph possible.

Sukanta Bhattacharjee
Bhargab B. Bhattacharya
Krishnendu Chakrabarty

List of Figures

List of Tables

List of Abbreviations

μTAS	Miniaturized Total Analysis System
ADSS	Approximate Digital Straight Line Segment
CAD	Computer-Aided Design
CF	Concentration Factor
CFMB	Continuous Flow-Based Microfluidic Biochip
CoDOS	Common Dilution Operation Sharing
DMFB	Digital Microfluidic Biochips
DMRW	Dilution and Mixing with Reduced Wastage
DSS	Digital Straight Line Segment
EWOD	Electrowetting-On-Dielectric
Ex-CoDOS	Extended Common Dilution Operation Sharing
FloSPA-D	Flow-Based Sample Preaparation Algorithm for Dilution
FloSPA-EM	Flow-Based Sample Preaparation Algorithm for Enhanced Mixing
FloSPA-M	Flow-Based Sample Preaparation Algorithm for Mixing
GDA	Generalized Dilution Algorithm
GMA	Generalized Mixing Algorithm
GORMA	Graph-Based Optimal Reactant Minimization Algorithm
IC	Integrated Circuit
IDMA	Improved Dilution/Mixing Algorithm
ILP	Integer Linear Programming
LDT	Linear Dilution Tree
LoC	Lab-on-a-Chip
MDST	Multiple Droplets of Single Target
MEDA	Micro-Electrode Dot Array
MIC	Minimum Inhibitory Concentration
mLSI	Microfluidic Large-Scale Integration
MSE	Mean-Square-Error

MTC	Multiple Target Concentration
MTCS	Mixing Tree with Common Subtree
MTDG	Multiple Target Droplet Generation
NFOSPA	Network-Flow-Based Optimal Sample Preparation Algorithm
PBDA	Pruning-Based Dilution Algorithm
PCR	Polymerase Chain Reaction
PCV	Prime Concentration Value
PDMS	Polydimethylsiloxane
PoC	Point of Care
REMIA	Reactant Minimization Algorithm
RMA	Ratio-ed Mixing Algorithm
RSM	Reagent-Saving Mixing
SAT	Satisfiability
SMT	Satisfiability Modulo Theory
SMTLA	Satisfiability Modulo Theory over Linear Arithmetic
TPG	Tree Pruning and Grafting
VOSPA	Volume-Oriented Sample Preparation Algorithm
WARA	Waste Recycling Algorithm

1

Introduction

Microfluidic technology is one of the major driving forces that steers the miniaturization of laboratory-based biochemical protocols on a tiny chip. Rapid sample processing, usage of nano- or pico-liter-range sample volumes, and the capability of precise control of fluids in an assay have made microfluidic chips an attractive choice to replace traditional bench-top diagnostic procedures and associated bulky instruments with a miniaturized total analysis system (μTAS) [RIAM02] or lab-on-a-chip (LoC) [SFB14].

One of the great challenges in science and engineering today is to develop technologies to improve the community health facilities in the poorest regions of the world. Among the worlds total population, more than 1 billion people lack basic healthcare services [YEF⁺06]. Apart from public healthcare, the market revenue for in vitro diagnostics is expected to reach $74.65 billion by 2020, from the market value of $53.32 billion in 2013, at a compound annual growth rate of 5.34% during 2014–2020 [InV]. Microfluidic systems allow miniaturization and integration of complex laboratory operations into a tiny chip that provides inexpensive and reliable PoC diagnosis platform for HIV [CLC⁺11], diabetes [SHT⁺08], immunoassay [SPO⁺08], cardiovascular diseases [CTF⁺02], and hence, they are well suited to the medical and social contexts of the developing world. Recent advances in microfluidic technologies spearhead new perspectives of research in molecular biology, genetic engineering, cell studies [VCLBPT10], drug discovery [DM06, NGL⁺12], chemotaxis assays [MMVKI10], and DNA analysis [MGC⁺11, AHS⁺13]. LoC devices can now be conveniently integrated to smartphone-based platforms for data processing and communication with applications to public health monitoring and diagnostics [EOJ⁺14].

1.1 Basics of Microfluidic Biochips

Microfluidic biochips are broadly categorized into two types: continuous-flow-based and digital or droplet-based.

Flow-based Microfluidic Biochips

There are three major categories of continuous flow-based microfluidic biochips (CFMBs): (i) free-flowing chips, which are built with pre-designed and fixed network of only flow channels, and fluids are pumped in through the inlets using pressure gradients [FPM12, SIP16a, TGZ+17, DCJW01, SWJ08, LKA+09, WJWL10], (ii) segmented-flow based chips, where a carrier fluid is used to navigate discrete droplets [SBPH12, SBV11, LYDP17] through a network of channels, and (iii) valve-based chips, which are equipped with a network of flow channels, and pressure-driven micro-valves via control channels that enable controlled and metered fluid flow through the network [HQ03, AB14]. Because of the ease of programmability, valve-based CFMBs are preferred for algorithmic sample preparation. This technology allows hundreds of assays to be performed in parallel in an automated fashion. Figure 1.1 shows a valve-based CFMB platform used for DNA purification of bacterial cell culture [HSH+04].

A typical valve-based microfluidic device consists of two separate layers of elastomer (polydimethylsiloxane (PDMS)) called flow layer and control layer as shown in Figure 1.2. Soft-lithographic techniques are used to

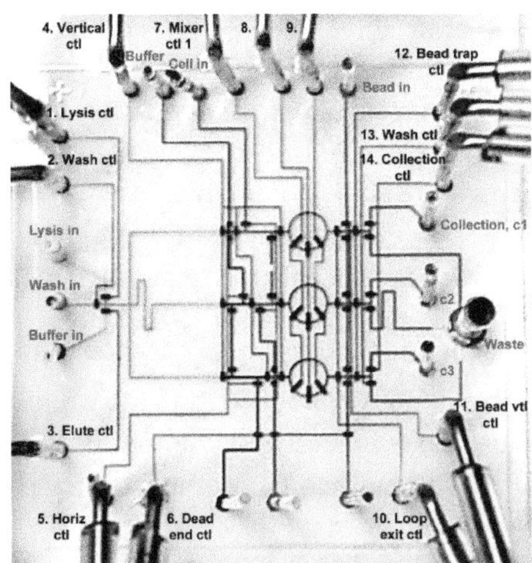

Figure 1.1 CFMB used for DNA purification from bacterial cell culture [HSH+04].

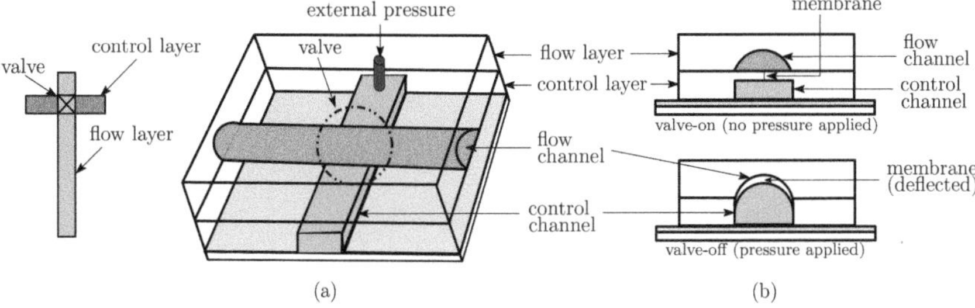

Figure 1.2 Schematic of a two-layer microfluidic device: (a) top-view, (b) cross-sectional view.

fabricate the network of microchannels in both flow and control layers. A flexible membrane/microvalve is formed at the overlapping area between the channels of two layers that can be used to control the movement of biochemical fluids in the flow layer. An external pressure source in the control layer is used to deflect membrane/valve deep into the flow channels to block fluidic flow. Channels in flow layers are also connected to an external pressure source. Recent technologies for microfluidic large-scale integration (mLSI) [HQ03, AB14] enable a CFMB to accommodate thousands of valves on a single chip. Based on these valve operations, many fluidic components can be constructed such as switches, elastomeric pumps [KLP+12], rotary mixers, and multiplexers/demultiplexers [MHR+10]. The movement of fluids through microfluidic channel can be controlled by configuring valves as switches (used for restricting/allowing flow in a channel depending on their closed/open states) to establish the desired flow path, as shown in Figure 1.3.

It is possible to build more complex fluidic modules by properly placing valves into the flow channel. For example, a rotary mixture is used in

$c_1 = c_2 = 1$ (external pressure released, valves open, flow passes)
$c_3 = c_4 = 0$ (external pressure applied, valves closed, flow blocked)

c_i: valve control, $c_i = 0/1$, valve v_i is closed/open

Figure 1.3 Controlling fluid movement through a microfluidic channel by configuring valves as switches.

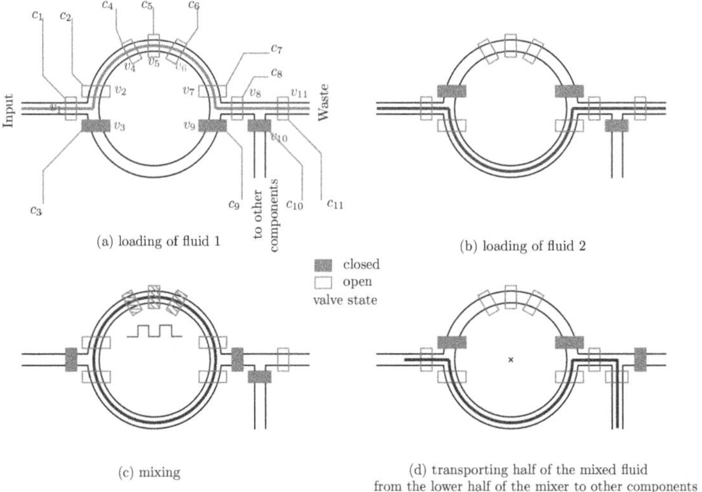

Figure 1.4 Schematic of rotary mixer and its operational phases for implementing (1:1) mixing model.

CFMB for mixing two or more input fluids in a desired ratio. Figure 1.4 shows a schematic diagram of a rotary mixer for mixing two fluids in (1:1) ratio and its operational phases.[1] The continuous-flow rotary mixer has five operational phases. In the first two phases, two input fluids are loaded into the upper and lower halves of the rotary mixer. Valve-states for loading input fluid into the upper half of the mixer are shown in Figure 1.4(a). After configuring valve-states, input fluid is pumped to fill the upper half of the mixer. Similarly, the other input fluid can be pumped to fill the lower half of the mixer (Figure 1.4(b)). After filling the mixer, valves v_1 and v_8 are closed and an actuation pattern is applied on $\{c_4, c_5, c_6\}$ to open (deflate) or close (compress) the valves $\{v_4, v_5, v_6\}$ in a sequence such that the liquid rotates through the circular flow channel in order to perform the desired mixing operation (Figure 1.4(c)). Finally, both halves of the mixed sample can be pushed out from the mixer to other components or storage cells. Moreover, precise metering of fluid sample can be carried out by transporting the sample between two valves that are located a fixed length apart. The detailed description of several other fluidic components along with their working principles are demonstrated in [MQ07, UTR$^+$06, MPP11].

[1]Control lines $c_1, c_2, \ldots, c_9$ are used for opening or closing of valves $v_1, v_2, \ldots, v_9$, respectively. Note that $\{c_1, c_2, c_3\}$ and $\{c_7, c_8, c_9\}$ are used for loading and unloading fluids, whereas $\{c_4, c_5, c_6\}$ control pumping operation of the mixer.

Digital Microfluidic Biochips

Digital microfluidic biochips (DMFBs) [CNFW12] use electrical actuation to manipulate (dispensing, navigation, merging, mixing, splitting, washing, sensing) discrete droplets of nanoliter/picoliter volume of reactant fluids on a two-dimensional electrode array. The salient features of DMFB lie on the controllability of each individual droplet without the need for networks of channels, pumps, valves, or mixers. Hence, various fluidic operations can be performed anywhere on the chip in a reconfigurable manner. Nevertheless, reconfigurability allows the same chip to be used for multiple applications, which helps to build a general-purpose programmable microfluidic platform.

The schematic diaram of a DMFB is shown in Figure 1.5. The top-view of a DMFB is shown in Figure 1.5(a). The cross-sectional view of a DMFB detection cell is shown in Figure 1.5(b), representing droplet of polarizable and/or conductive liquid that is sandwiched between two layers of hydrophobic insulator. On the upper layer, there is a single plate of continuous ground electrode, while underneath the bottom layer, there is an array of electrode-plates. The gap between two parallel plates is usually filled with silicone oil, which acts as a filler fluid that prevents droplet evaporation and reduces surface contamination. Other peripheral devices such as optical/electronic detectors and dispensing ports are also integrated with DMFB.

Among several technologies for manipulating fluid droplets, electrowetting-on-dielectric (EWOD) technology [MB05] is widely used for DMFBs. EWOD technology uses electric field to modify the wetting behavior of a polarizable and/or conductive liquid droplet in contact with a hydrophobic, insulated electrode. Application of voltage between the liquid and the electrode creates in an electric field across the insulator, which, in turn, lowers the interfacial tension between the liquid and the insulator surface

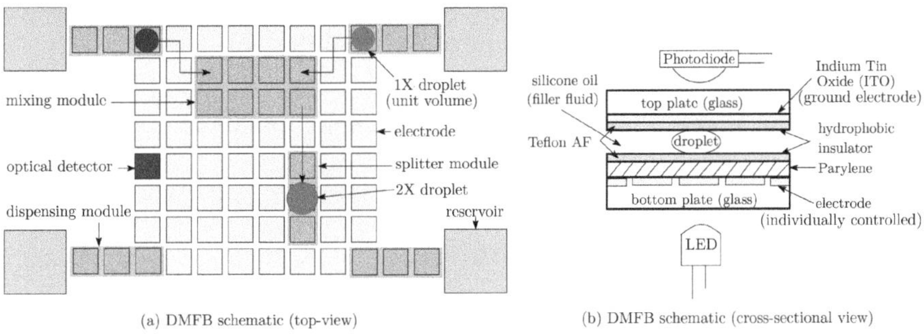

Figure 1.5 DMFB schematic (a) top view and (b) cross-sectional view.

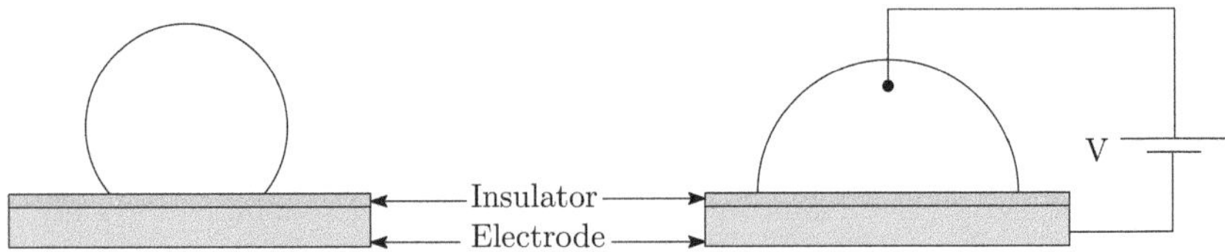

Figure 1.6 Electrowetting-on-dielectric (EWOD) effect.

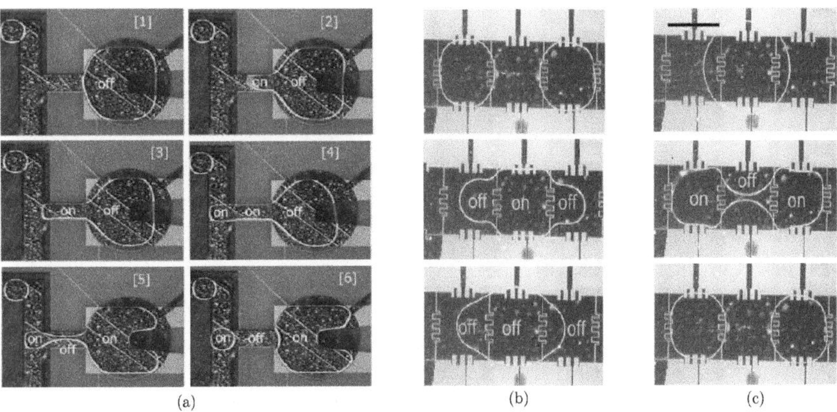

Figure 1.7 DMFB fluidic operations (a) droplet dispensing [SPF04], (b) droplet mixing, and (c) splitting [CMK03] under the (1:1) mix/split model.

according to the Lippman–Young equation [MB05]. Figure 1.6 illustrates the EWOD-effect on a droplet, which is initially at rest on a hydrophobic insulated electrode. Application of voltage (V) reduces the solid–liquid interfacial tension, resulting in increased wetting of the surface by the droplet. The application of certain voltage pattern to a series of adjacent electrodes creates an interfacial tension gradient, which can be used to manipulate droplets. The detailed description of DMFB operations and their underlying principles are elaborated in [Fai07, SPF04, CMK03, PFS00]. Figure 1.7(a) shows the dispense operation of a droplet from an on-chip reservoir [SPF04]. Figure 1.7(b) and Figure 1.7(c) show the (1:1) droplet mixing and splitting operations on a DMFB [CMK03], respectively.

1.2 Design Automation of Microfluidic Biochips

Ongoing research activities in microfluidics are highly multi-disciplinary in nature. Engineers working in this field need to collaborate with biologists and clinicians for developing new applications and ideas to be implemented using microfluidic toolbox. Recent years have seen considerable development in the computer-aided design (CAD) automation in microfluidic biochips [CS07, PAC15, ILC15, Ho14, TLHS15, TLL$^+$16, TLSH15, YHC15]. These CAD tools act as interface between biologists and engineers that allow the users to focus on the development of bioassays, leaving aside chip optimization and implementation details to design automation tools. It is expected that an automated design flow will transform the biochip research and use, in the same way as design automation revolutionized the IC design flow in the 1980s and the 1990s [HZC10, CS07]. This approach is therefore especially aligned with the vision of functional diversification and "More than Moore" as articulated in the International Technology Roadmap for Semiconductors 2007 [ITR], which highlights Medical as being a System Driver for the future.

A synthesis tool [GB14, GCW$^+$15] typically executes a series of transformations for implementing an input protocol description on a target architecture, as shown in Figure 1.8. At first, the behavioral description of a bioassay is represented using a sequencing graph [CS07] in which nodes denote fluid operations and the edges represent dependencies. This is followed by a sequence of synthesis steps, which include resource binding, scheduling, module placement, and droplet routing. Finally, depending on the design constraints (e.g., array area, assay completion time, and resource constraints), a detailed layout of the DMFB along with the sequence of actuation steps are generated [CS07].

1.3 Sample Preparation with Microfluidic Biochips

Sample preparation is an indispensable step in almost all biochemical protocols for mixing two or more biochemical reagents in a given volumetric ratio. Dilution is a special case of sample preparation, where only two input reagents (commonly known as sample and buffer) are mixed in a desired volumetric ratio. A real-life biochemical laboratory protocol often requires mixing of several reagent fluids. For example, a sample of 70% ethanol is required in Glucose Tolerance Test in mice, and in E.coli Genomic DNA Extraction [bio]. Also, 95% ethanol is required for total RNA extraction from worms [bio]. A sample of 10% Fetal Bovine Serum (FBS) is required for in vitro culture of human Peripheral Blood Mononuclear Cells (PBMCs) [bio]. In the polymerase chain reaction (PCR) used for DNA amplification, a master-mixture of seven fluids (reactant buffer, dNTPs, forward primer, reverse primer, DNA template, optimase, and water)

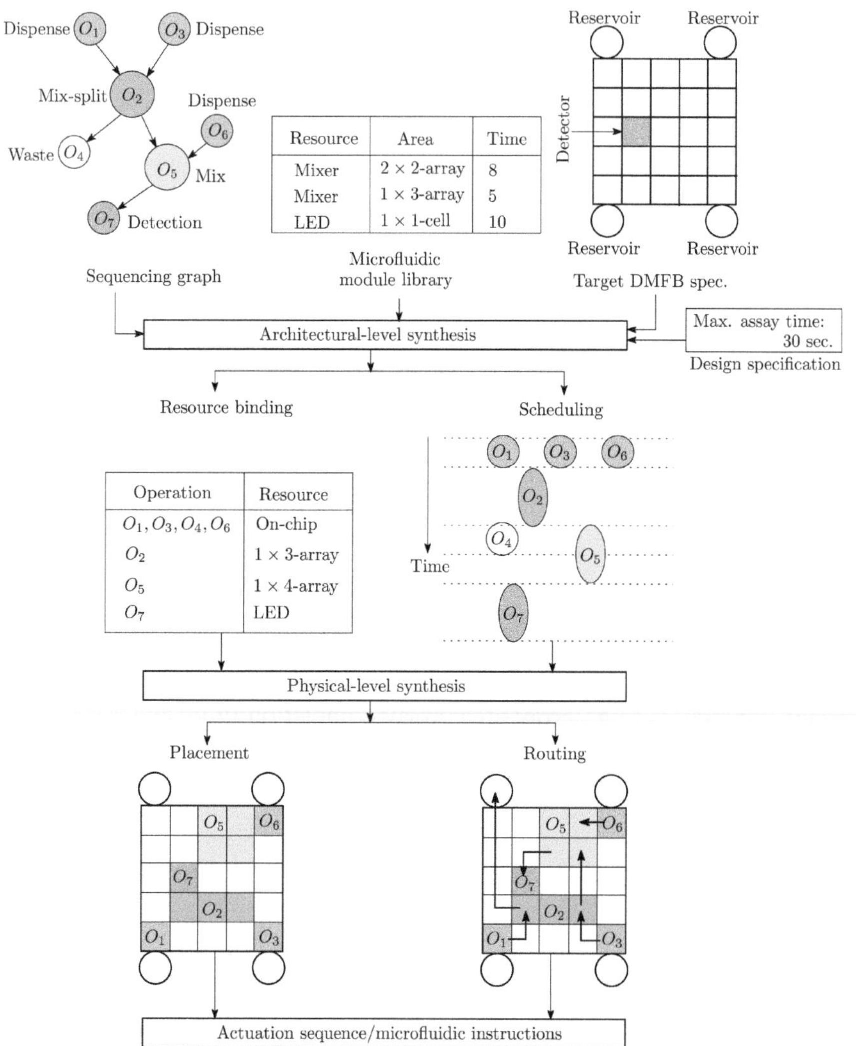

Figure 1.8 DMFB synthesis flow.

is required with a volumetric ratio $\{10\% : 8\% : 0.8\% : 0.8\% : 1\% : 1\% : 78.4\%\}$ [PCRa]. Moreover, a mixture of three reagents is required in "One-step miniprep method" [RCK+15b] and also for the preparation of plasmid DNA by alkaline lysis with SDS-minipreparation [HHC14].

Sample preparation can be performed both on-chip or outside the chip as a pre-processing step. Off-chip sample preparation not only elongates the overall assay completion time but also increases sample-preparation cost.

In molecular diagnosis, 90% of cost and 95% of time is associated with sample collection, transportation, and preparation [GV04]. Microfluidics offers the promise of a "lab-on-a-chip" system that can individually control picoliter-scale quantities of fluids, with integrated support for operations such as mixing and storage, and hence provides a low-cost and reliable solution for on-chip sample preparation. Microfluidics-based sample-preparation algorithms usually perform a sequence of mixing operations following an on-chip mixing primitive (abstracted as mixing model) to obtain a target mixture using a microfluidic chip. Although the capabilities of on-chip mixers may vary from one chip to another, a simple (1:1) mixing model has been found to be suitable for both continuous-flow and droplet-based architectures. A (1:1) mixing operation combines two units of fluids in equal proportions, producing two units of the mixture. This may be followed by a balanced splitting that produces two droplets, each having unit volume. Figure 1.9 and Figure 1.10 show the (1:1) mixing model implemented on DMFB and CFMB platforms, respectively.

Basics of Sample Preparation

A mixture of k reagents $R_1, R_2, \ldots, R_k$ is denoted as $\mathcal{M} = \{\langle R_1, c_1\rangle, \langle R_2, c_2\rangle, \ldots, \langle R_k, c_k\rangle\}$, where $\sum_{i=1}^{k} c_i = 1$ (validity condition) and $0 \leq c_i \leq 1$ for $i = 1, 2, \ldots, k$. In other words, $R_1, R_2, \ldots, R_i, \ldots, R_k$ are mixed with a ratio of $\{c_1 : c_2 : \cdots : c_i : \cdots : c_k\}$, where c_i denotes the concentration factor (*CF*) of R_i. The condition $\sum_{i=1}^{k} c_i = 1$ [TUTA08] ensures the validity of a mixing ratio. Note that the *CF* of a pure (100%) reagent R_i is assumed to be 1, and the *CF* of neutral buffer (0%) is assumed to be 0. Thus, the value of each c_i is normalized with respect to 1, i.e., the sum

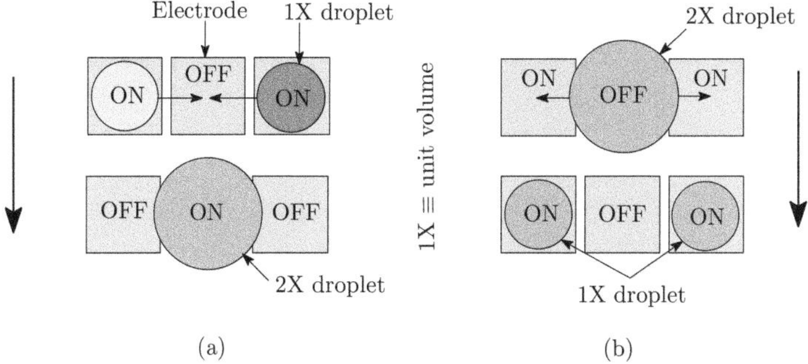

Figure 1.9 Schematic of (1:1) (a) mixing and (b) splitting of droplets on a DMFB.

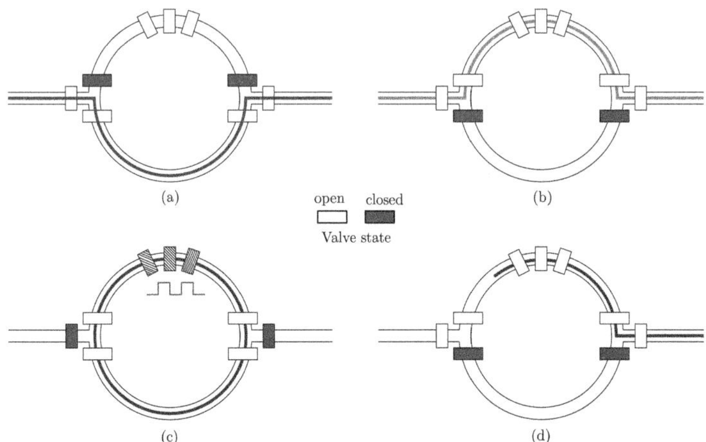

Figure 1.10 Schematic of (1:1) mixing on a CFMB, where (a) and (b) show the loading of equal amount of input reagents into a rotary mixer, (c) shows the mixing operation, and (d) shows the transportation of unit-sized sample after mixing.

of all c_i's should be equal to 1. Because of the inherent mixing models used for microfluidic implementation of mixers, each c_i is required to be approximated as a special form depending on the mixing model and user-defined error-tolerance limit ϵ in *CF*, $0 \leq \epsilon < 1$. In (1:1) mixing model, the set of reachable mixing ratios are of the form $\mathcal{M} = \{\langle R_1, c_1 = \frac{x_1}{2^d}\rangle, \langle R_2, c_2 = \frac{x_2}{2^d}\rangle, \ldots, \langle R_k, c_k = \frac{x_k}{2^d}\rangle\}$, where $\sum_{i=1}^{k} x_i = 2^d$ and $x_1, x_2, \ldots, x_k, d \in \mathbb{N}$ [TUTA08]. Note that d is determined by the desired accuracy of approximation, i.e., error-tolerance limit ϵ in *CF*, $0 \leq \epsilon < 1$, which is user-specified. A lower value of ϵ denotes higher accuracy in *CF*. Formally, assuming the (1:1) mixing model, for a given mixture $\mathcal{M} = \{\langle R_1, c_1\rangle, \langle R_2, c_2\rangle, \ldots, \langle R_k, c_k\rangle\}$ and error-tolerance limit ϵ in *CF*, we need to choose the minimum value of $d \in \mathbb{N}$ such that each c_i in mixture $\mathcal{M}$ is approximated as $\frac{x_i}{2^d}$, where $x_i \in \mathbb{N}$, subject to $\max_i\{|c_i - \frac{x_i}{2^d}|\} < \epsilon$ and $\sum_{i=1}^{k} x_i = 2^d$ [TUTA08]. Hence, for the input mixture $\mathcal{M}$, the approximated target ratio becomes $\{R_1 : R_2 : \cdots : R_k = x_1 : x_2 : \cdots x_k\}$, which is reachable under (1:1) mixing model satisfying user-defined error-tolerance limit ϵ in *CF*. Sample preparation algorithms start with an approximated target ratio and produce a sequence of mixing steps (commonly represented using sequencing graph [CS07]) for reaching the target ratio staring from raw input reagents. Figure 1.11 shows the overview of sample preparation on a microfluidic platform.

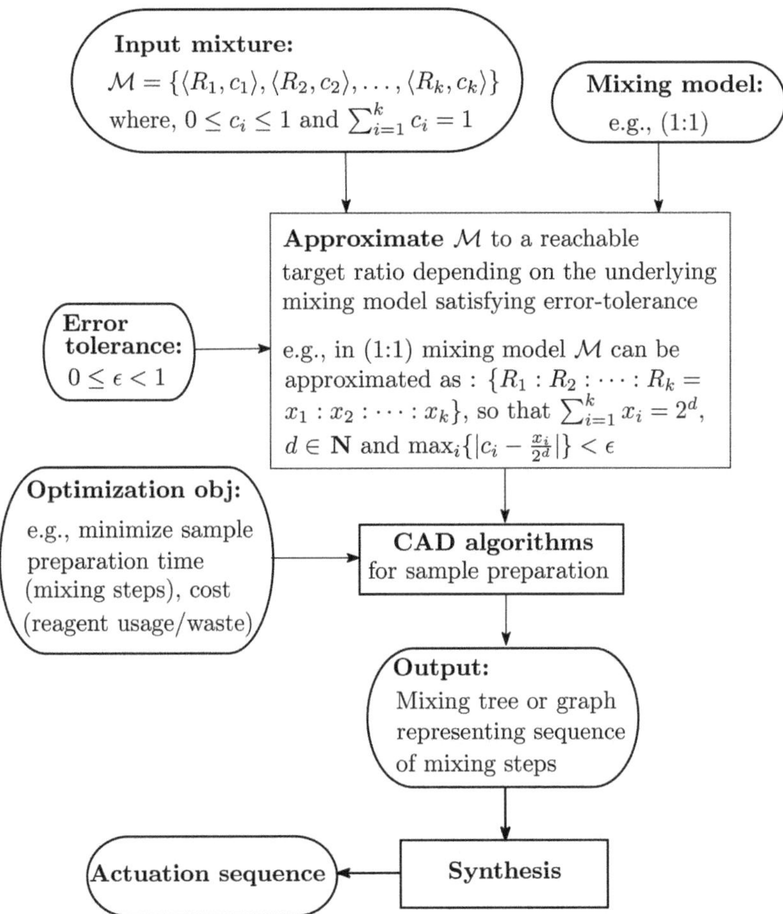

Figure 1.11 Overview of sample preparation on microfluidic biochips.

Several sample-preparation algorithms have been proposed for mixing two (dilution) or more input reagents on both DMFB and CFMB platforms. Most of DMFB-based algorithms [HHC14, TUTA08, RBC10, RBC11, LCH15, CLH13, HLL13, RCK+15b, LCLH13, KRC+13, DYHH13] use (1:1) mixing model. However, few algorithms refer to sample preparation on CFMB using multiple mixing models [LSH15, HLH15]. Thies *et al.* proposed a mixing algorithm *MinMix* [TUTA08] for generating a target ratio of input reagents on a microfluidic biochip supporting (1:1) mixing model. In this method, the target ratio of k input reagents $\mathcal{M} = \{\langle R_1, c_1\rangle, \langle R_2, c_2\rangle, \ldots, \langle R_k, c_k\rangle\}$ is approximated as $\{R_1 : R_2 : \cdots : R_k = x_1 : x_2 : \cdots : x_k\}$, where $\sum_{i=1}^{k} x_i = 2^d$, by choosing suitable $d \in \mathbb{N}$ depending on the user-defined error-tolerance limit $0 \leq \epsilon < 1$. Next, each x_i is represented as a d-bit binary string and scanned from right to left to construct a mixing tree in a bottom-up fashion. As an example, consider the mixture of three input reagents $\mathcal{M} = \{\langle R_1, 0.22\rangle, \langle R_2, 0.44\rangle, \langle R_3, 0.34\rangle\}$ required in preparation of plasmid DNS by alkaline lysis with SDS-minipreparation [HHC14]. For the user-defined error-tolerance limit $\epsilon = 0.005$, $\mathcal{M}$ is approximated as $\{R_1 : R_2 : R_3 = 7 : 14 : 11\}$ by choosing $d = 5$, since $\max\{|0.22 - \frac{7}{2^5}|, |0.44 - \frac{14}{2^5}|, |0.34 - \frac{11}{2^5}|\} = \max\{0.001, 0.002, 0.003\} = 0.003 < \epsilon$. The mixing tree[2] for the approximated target ratio $\{7 : 14 : 11\}$ is shown in Figure 1.12. Note that the depth of the mixing tree is d. Moreover, the number of mixing steps and waste droplets are denoted as n_m and n_w, respectively, and n_{r_i} denotes the number of droplets of reagent R_i, for $i = 1, 2, 3$. In the case of dilution, the algorithm *twoWayMix* (*MinMix* with two input reagents) generates a dilution tree with minimum number of mixing steps.

In automatic sample preparation, it is desirable to reduce the sample-preparation time and cost. Note that the number of mixing steps is proportional to sample-preparation time and the amount of waste generated during

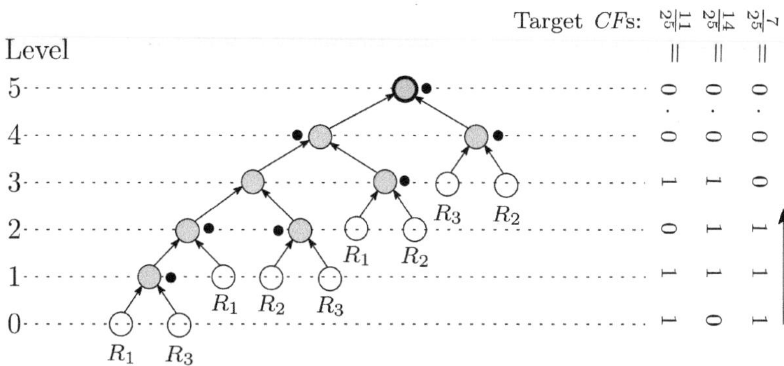

Figure 1.12 Mixing tree for the target ratio $\{R_1 : R_2 : R_3 = 7 : 14 : 11\}$ produced by *MinMix* [TUTA08].

[2]A dilution or mixing tree is a tree where each leaf node represents one unit volume of an input fluid, and a non-leaf node denotes a mixing/splitting operation of inputs and/or intermediate fluids. In the case of (1:1) mixing model, the underlying dilution or mixing tree is a binary tree.

sample preparation corresponds to cost. Moreover, it is essential to reduce the usage of expensive/precious reagent, and hence, reactant minimization also arises as an important issue during sample preparation.

In many real-life applications, a variety of concentrations for the same sample are often required [bio, MRB$^+$14, CHDW12, SHK10], e.g., in bacterial susceptibility tests, samples and reagents are required in multiple concentration (or dilution) factors, satisfying certain "gradient patterns" such as linear, exponential, or parabolic. A straightforward adaptation of the existing single-target-based methods requires the dilution steps for each target to be executed separately, thereby leading to a large number of mixing steps (i.e., long execution time) and increased wastage of precious samples and expensive reagents. Hence, it is a challenge to carry out on-chip sample preparation for achieving the desired target *CF*s minimizing the number of mixing and waste. Sample preparation becomes even more challenging on CFMB because of the availability of on-chip multiple mixing models. Note that multiple mixing model provides more flexibility in sharing of intermediate *CF*s, thereby reducing the number of mixing steps compared to (1:1) mixing graph. Efficient sample preparation on CFMB, considering generalized mixing model, was posed as an open problem earlier [TUTA08].

In this research monograph, we discuss various techniques for efficient generation of multiple target-*CF*s of a sample and for producing different types of dilution gradients using DMFBs. Additionally, for CFMBs, we describe algorithmic solutions to sample preparation assuming availability of multiple mixing models, and also, under the constraint of limited availability of on-chip storage.

1.4 Organization of the Book

The book is organized as follows. Chapter 2 provides a comprehensive survey of existing sample-preparation techniques. Chapter 3 discusses the problem of generating multiple *CF*s with DMFBs. Algorithms for generating various types of dilution gradients are presented in Chapter 4. Chapter 5 describes a method for concentration-resilient ratio selection that can be used as a preprocessing step for selecting a suitable mixing ratio before invoking sample-preparation algorithms. Dilution and mixing algorithms under the generalized mixing model supported by flow-based microfluidic biochips are discussed in Chapter 6. Sample preparation with limited on-chip storage is described in Chapter 7. Finally, the book ends with a brief conclusion and future research directions in Chapter 8.

2

Sample Preparation with Microfluidic Biochips: A Review

Sample preparation is the process of mixing two or more biochemical fluidic reagents in a given volumetric ratio through a sequence of mixing operations. In sample preparation, a mixture of k input reagents $\mathcal{M} = \{\langle R_1, c_1\rangle, \langle R_2, c_2\rangle, \ldots, \langle R_k, c_k\rangle\}$,[1] where $\sum_{i=1}^{k} c_i = 1$ and $0 \leq c_i \leq 1$ for $i = 1, 2, \ldots, k$, is approximated as $\{R_1 : R_2 : \cdots : R_k = x_1 : x_2 : \cdots : x_k\}$, where $\sum_{i=1}^{k} x_i = N^d, N, d \in \mathbb{N}$. The value of N is determined by the mixing model supported by the microfluidic mixer (e.g., $N = 2$ for (1:1) mixing model), and d is determined by the desired accuracy of approximation (error-tolerance limit in *CF*, $0 \leq \epsilon < 1$), which is user-specified. A lower value of ϵ denotes higher accuracy in *CF*. The detailed description of ratio approximation can be found in Chapter 1. After approximating the input mixture $\mathcal{M}$, sample preparation algorithms construct a mixing tree or graph depending on the optimization objectives such as minimization of mixing steps, waste generation, input reagent consumptions, and so on.

Most of the existing sample preparation algorithms are based on (1:1) mixing model as it is easy to implement in both DMFB and CFMB platforms. In (1:1) mixing model, two fluids, having ratio of input fluids $\{R_1 : R_2 : \cdots : R_k = x_1 : x_2 : \cdots : x_k\}$ and $\{R_1 : R_2 : \cdots : R_k = y_1 : y_2 : \cdots : y_k\}$ respectively, are mixed into equal volumetric ratio, and after mixing, it is split into two equal parts, each having concentration ratio $\{R_1 : R_2 : \cdots : R_k = \frac{x_1+y_1}{2} : \frac{x_2+y_2}{2} : \cdots : \frac{x_k+y_k}{2}\}$. In the case of dilution, where two input reagents (commonly known as sample and buffer) are mixed, the approximated ratio $\{\text{sample} : \text{buffer} = x : y\}$ can alternatively be represented as the *CF* of sample, i.e., $\frac{x}{x+y}$ or simply by $CF = \frac{x}{2^d}$,

[1] c_i denotes the concentration factor (*CF*) of R_i.

Table 2.1 Symbols/notations used for sample preparation algorithms

	Symbol/notation	Meaning
1.	$\mathcal{M} = \{\langle R_1, c_1\rangle, \langle R_2, c_2\rangle, \ldots, \langle R_k, c_k\rangle\}$, where $\sum_{i=1}^{k} c_i = 1$ and $0 \leq c_i \leq 1$ for $i = 1, 2, \ldots, k$	A mixture of k input reagents. In the special case of dilution, $k = 2$
2.	c_i	Concentration factor (*CF*) of R_i
3.	$0 \leq \epsilon < 1$	User-defined error-tolerance limit in *CF*
4.	$\{R_1 : R_2 : \cdots : R_k = x_1 : x_2 : \cdots : x_k\}$, where $\sum_{i=1}^{k} x_i = N^d, N, d \in \mathbb{N}$	Approximated mixing ratio where *CF* of R_i is $\frac{x_i}{N^d}$. In case of (1:1) mixing model, N is set to 2
5.	d	Depth of a mixing tree for a target ratio
6.	n_m	Number of mixing
7.	n_s (n_b)	Number of sample (buffer) droplets or segments
8.	n_{r_i}	Number of segments filled up with reagent R_i
9.	n_w	Number of waste droplets or segments
10.	n_r	Total number of segments filled up with input reagents

where $x + y = 2^d$. Table 2.1 summarizes the symbols/notations used in sample preparation algorithms.

Thies *et al.* [TUTA08] proposed a mixture preparation algorithm (*Min-Mix*) for mixing two or more input fluids in a given ratio $\{R_1 : R_2 : \cdots : R_k = x_1 : x_2 : \cdots : x_k\}$, where $\sum_{i=1}^{k} x_i = 2^d$, assuming (1:1) mixing model. Note that the depth of the mixing tree turns out to be d, as we have chosen minimum d for approximating an input mixture $\mathcal{M}$ to a reachable target ratio $\{x_1 : x_2 : \cdots : x_k\}$ depending on the accuracy ϵ. In *MinMix* algorithm, each x_i is represented as a d-bit binary number. After that, the bits of the (0,1) matrix of size $k \times d$ are scanned column-wise from right to left to construct a binary mixing tree in a bottom-up fashion. Figure 2.1(a) shows a mixing tree for the approximated target ratio $\{R_1 : R_2 : R_3 = 6 : 7 : 3\}$. Algorithm *MinMix* starts scanning the rightmost k bits of binary representation of the portion of each R_i in the input mixture for determining the input reagent required at level-0 in the mixing tree. Each non-zero bit represents an input reagent at level-0. As shown in Figure 2.1(a), the rightmost bits in the 4-bit binary representation of R_1, R_2, R_3 are $0, 1, 1$, respectively; hence, at level-0 a reagent droplet of each R_2 and R_3 is required. Analogously, *MinMix* scans

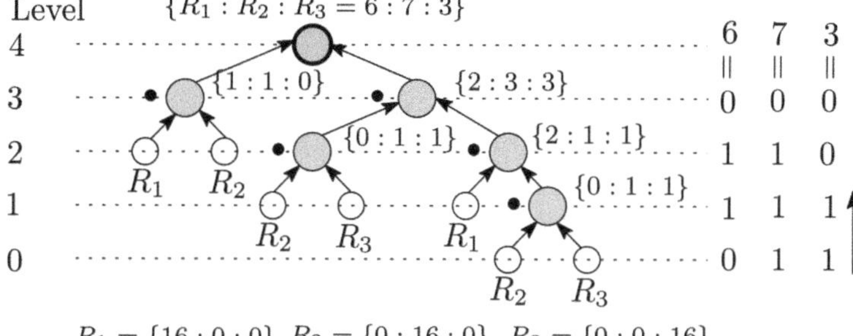

Figure 2.1 Mixture preparation using *MinMix* [TUTA08]: binary tree representation.

the next k-bit column for determining the input reagents needed at the next level. Finally, a binary mixing tree is constructed after level-wise merging of two leaf nodes (or, one leaf node and one non-leaf nodes, or two non-leaf nodes) indicating the required mix-split steps for the corresponding input fluid or intermediate droplets. In a mixing tree, each leaf node corresponds to a unit-sized droplet (1X-volume) and an internal node represents a (1:1) mix-split operation, where the actual physical volume of a 1X-size droplet is determined by the electrode size, actuation voltages, and the underlying microfluidic technology.

2.1 Dilution Algorithms for DMFB

In this section, we provide a comprehensive description of the state-of-the-art sample preparation methods for dilution.

2.1.1 Single-target Dilution Algorithms

In the special case of dilution, where only two input reagents are mixed, *twoWayMix*[2] [TUTA08] generates a mixing tree that requires the minimum number of mixing operations. The mixing tree produced by *twoWayMix* becomes skewed in shape (Figure 2.2(a)). Note that *twoWayMix* does not require to store any intermediate droplet; only the current droplet is mixed with the input droplet in each mixing step except the first mixing step,

[2]*MinMix* for $k = 2$ i.e., mixing two input reagents only.

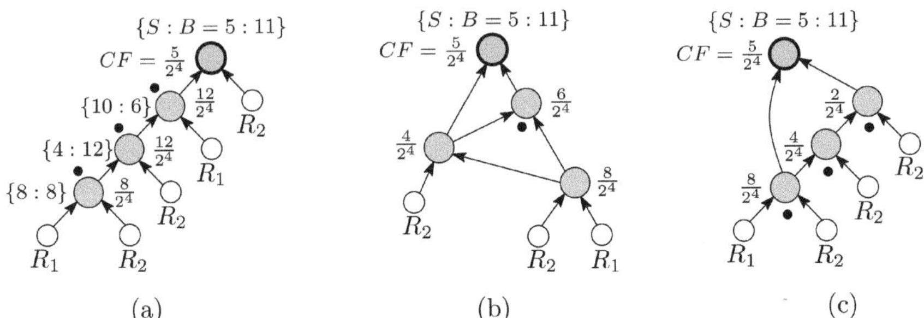

Figure 2.2 Dilution of a sample using (a) *twoWayMix* [TUTA08], (b) *DMRW* [RBC10], and (c) *REMIA* [LCH15].

where only two input droplets are mixed. However, this method produces one waste droplet at each mixing step except the last one, and hence, it leads to high input consumption and waste management overhead. A mixing tree generated using *twoWayMix* is shown in Figure 2.2(a) for the target $CF = \frac{5}{2^4}$ or the target ratio $\{\text{sample} : \text{buffer} = 5 : 11\}$. Note that the number of (1:1)-mixing step (n_m), waste droplets (n_w), input sample droplets (n_s), and buffer droplets (n_b) are $4, 3, 2$, and 3, respectively.

Another algorithm called 'Dilution and mixing with reduced wastage' (*DMRW*) [RBC10] shares intermediate droplets for reducing waste droplets. It uses a binary search technique for generating a target concentration. Algorithm *DMRW* starts by setting sample and buffer as upper and lower bounds, respectively, and in each iteration, the lower (upper) bound is updated with the average of the two if the target is larger (smaller) than the average. The process terminates until the target is reached. As *DMRW* shares intermediate droplets for minimizing waste droplets, some extra storage space is required for storing intermediate droplets. In order to facilitate its implementation, specialized rotary mixer based on $(m : m)$, $m \geq 1$, mixing model is used for speeding dilution time. Figure 2.2(b) shows the dilution tree for $CF = \frac{5}{2^4}$ generated with *DMRW*. Note that for this target *CF*, *DMRW* generates only one waste droplet. An improved version of *DMRW*, called 'improved dilution/mixing algorithm' (*IDMA*) [RBC11] further improves the performance of *DMRW*.

Another important objective of sample preparation is to minimize the overall reactant usage. There are certain circumstances, e.g., where a blood sample from a premature baby or a DNA sample collected from a crime scene is needed, it is essential to reduce the demand of such scarcely available reagent. Huang *et al.* proposed a reactant-minimization algorithm (*REMIA*) [LCH15] for diluting a sample to the desired concentration factor. For a given target concentration, *REMIA* builds a mixing tree in which every leaf node indicates a prime concentration value (PCV). Note that a concentration value that can be obtained from a series of exponential dilutions [HLC12] starting from a raw reactant droplet is called a PCV. Finally, the PCV values of the the leaf nodes are produced through a series of exponential dilution operations with minimal usage of expensive reactant. Figure 2.2(c) shows the mixing graph for the target $CF = \frac{5}{2^4}$ using *REMIA*, where two PCV values ($\frac{8}{2^4}$, $\frac{2}{2^4}$) are used for generating $\frac{5}{2^4}$. Moreover, a series of three exponential dilution steps are required for generating three PCV values ($\frac{8}{2^4}$, $\frac{4}{2^4}$, $\frac{2}{2^4}$), which are obtained by repeatedly diluting R_1 with R_2. Another reactant-minimization algorithm, namely ‘graph-based optimal reactant minimization algorithm’ (*GORMA*) [CLH13], exhaustively checks all possible dilution solutions for a given target *CF* and finds the one with the least reactant-usage.

Roy *et al.* proposed a generalized dilution algorithm (*GDA*) [RBGC14b] for generating a target concentration factor (*CF*) using a set of input stock solutions. Given a supply of a sample with arbitrary *CF*s (each diluted with the same buffer), *GDA* aims to produce a desired dilution of sample using the fewest (1:1) mix-split steps. However, in the case of preparing a mixture of three or more reagents, Roy *et al.* derived a necessary and sufficient condition for obtaining a target ratio from a supply of $N(\geq\ 3)$ fluids, each provided with arbitrary *CF* values. A generalized mixing algorithm (*GMA*) [RBGC14b] is also proposed for generating the desired target ratio of N reagents using (1:1) mix-split steps, subject to the satisfaction of certain reachability conditions.

2.1.2 Multiple-target Dilution Algorithms

Many bioassays often require multiple concentration values of the same reagent, and implementing them efficiently on a digital microfluidic biochip is a challenge. Although multiple target sample preparation can be easily performed with the execution of the dilution steps for each target separately, this process requires a large number of mix-split steps and increases wastage of precious reagents. Bhattacharjee *et al.* proposed a pruning-based dilution algorithm (*PBDA*) [BBB14] for generating multiple dilutions of a given sample. Algorithm *PBDA* reuses the intermediate waste droplets (droplet sharing), provided each target concentration is generated using a single-target dilution algorithm *twoWayMix*. Additionally, an efficient search technique is used that selects in a cost-effective fashion, a suitable subset of intermediate droplets, which can be best reused (droplet replacement) for further mixing and waste minimization. Huang *et al.* proposed a multiple-target dilution algorithm, namely 'waste recycling algorithm' (*WARA*) [HLL13], which also uses droplet sharing and replacement technique for minimizing reactant-usage. *WARA* builds upon an earlier single-target sample preparation algorithm (*REMIA*) [LCH15] to create a reactant-minimized mixing tree for each target concentration. Next, it attempts to maximize droplet sharing and waste recycling among the sub-trees for reactant and waste minimization. Mitra *et al.* proposed an algorithmic solution to the multiple-target dilution problem, known as 'multiple target concentration' (*MTC*) [MRB+14], without using any on-chip storage for intermediate droplet, while optimizing the number of mixing steps. A DMFB platform for implementing multiple target dilution using *MTC* is also very easy to construct and the cost is significantly lower compared to those required by other multiple target dilution algorithms. Dinh *et al.* proposed a network-flow-based optimal sample preparation algorithm (*NFOSPA*) [DYHH13] for generating multiple-target multiple-demand dilution of an input sample. The proposed integer linear programming (ILP) model is flexible enough to allow the trade-off for minimizing the cost of sample or buffer. However, *NFOSPA* suffers from high computational overhead due to the large number of variables and constraints that appear in the underlying ILP modeling of minimum-cost integer flow problem [MS09].

2.1.3 Generation of Dilution Gradients

In many biological assays, e.g., in bacterial susceptibility tests, some reagents are required in multiple concentration (or dilution) factors, satisfying certain "gradient" patterns such as linear, exponential, or parabolic. Existing multiple-target dilution algorithms can be used to generate any gradient pattern. However, all existing algorithms fail to optimize cost or performance because none of them could utilize the embedded patterns hidden in the gradient profile. Bhattacharjee *et al.* presented an algorithm to generate an arbitrary linear gradient [BBH+13b], with minimum wastage, a subset of exponential gradients [BBH+13a], and complex-shaped gradients [BBH+ar] (e.g., exponential, parabolic, and sinusoidal) while satisfying the required accuracy in concentration factors. The proposed algorithms utilize the underlying combinatorial properties of a linear/exponential gradient for generating the target set.

2.2 Mixing Algorithms for DMFB

A number of sample preparation algorithms for mixing three or more input reagents based on (1:1) mixing model are also available in the literature. Roy *et al.* proposed a mixing algorithm, namely 'Ratio-ed mixing algorithm' (*RMA*) [RCK+15b], based on a number-partitioning technique that determines a layout-aware mixing tree corresponding to a given target ratio of several fluids. For a desired target ratio, RMA determines a mixing tree with longer sub-sequences of mixing steps with a small number of distinct fluids (called "dilution subtrees"[3]). Figure 2.3(a) shows a mixing tree for the target ratio $\{R_1 : R_2 : R_3 = 6 : 7 : 3\}$ generated by *RMA*, having two dilution subtrees. On-chip implementation of the mixing graph produced by *RMA* is convenient because its planer embedding requires fewer number of crossovers among droplet-routing paths as well as shorter reservoir-to-mixer transportation distance. The algorithm named 'mixing tree with common subtrees' (*MTCS*) [KRC+13] algorithm constructs a mixing tree for a target ratio by sharing the common subtrees produced by the *MinMix* [TUTA08] tree. *MTCS* permutes the leaf nodes and/or shifts the level of a leaf node of the *MinMix* tree for identifying common subtrees therein and utilizes unused

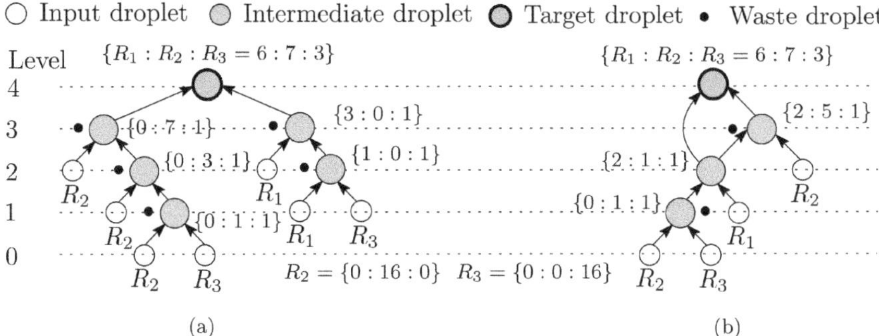

Figure 2.3 Mixture preparation using (a) *RMA* [RCK+15b] and (b) *MTCS* [KRC+13].

[3]A dilution subtree is a mixing tree in which the number of distinct leaf nodes is exactly two.

intermediate droplets in order to replace a common subtree. Figure 2.3(b) shows the mixing tree generated by *MTCS* starting from the *MinMix* tree, shown in Figure 2.1. Liu *et al.* proposed another mixing algorithm for single and multiple target mixture preparation, referred to as 'common dilution operation sharing' algorithm (*CoDOS* [LCLH13]) and 'extended *CoDOS*' (*Ex-CoDOS*) [LCLH13]. They first represent the given target concentration factors as a matrix and then identify rectangles therein, where each rectangle indicates a possible sharing of dilution operation that leads to reactant minimization. Hsieh *et al.* [HHC14] proposed 'reagent-saving mixing' algorithm (*RSM*) [HHC14] for determining a mixing graph for multiple target ratios of input fluids using a top-down decomposition technique. *RSM* decomposes the given input ratios to minimize the total number of input reagents and builds a composite mixing graph by reuse of intermediate waste droplets.

2.3 Droplet Streaming Algorithms

In many applications, a stream of target droplets with a desired target ratio may be required. For example, a bioassay may need to be repeated on samples for screening several patients in a point-of-care (PoC) diagnostic environment or an assay may require repeated execution to enhance reliability of test results. Roy *et al.* proposed two algorithms, namely 'multiple target droplet generation' (*MTDG*) [RBGC14a] and 'multiple droplets of single target' (*MDST*) [RKC$^+$14], for streaming droplets of two (dilution engine) or more (mixture-preparation engine) input reagents. *MTDG* (*MDST*) reuses the intermediate waste droplets produced by mixing tree/graph of single target dilution (mixture preparation) algorithms and thereby produces a stream of identical target droplets, instead of discarding the previously unused droplets as waste.

2.4 Dilution and Mixing Algorithms for CFMB

In a DMFB, two equal-volume droplets are mixed on the basis of the (1:1) mixing model. On the other hand, a CFMB supports multiple ratios as mixing models since a more versatile rotary mixer is available. In Ring-M, the first two segments are of equal size, while the volumetric ratio of the last $N-1$ segments, $1 : 2^1 : 2^2 : \cdots : 2^{M-1}$, forms a geometric sequence with a common ratio of 2. However, a Mixer-N is divided into N equal-length segments, where each segment can be filled with a fluid and it provides easier

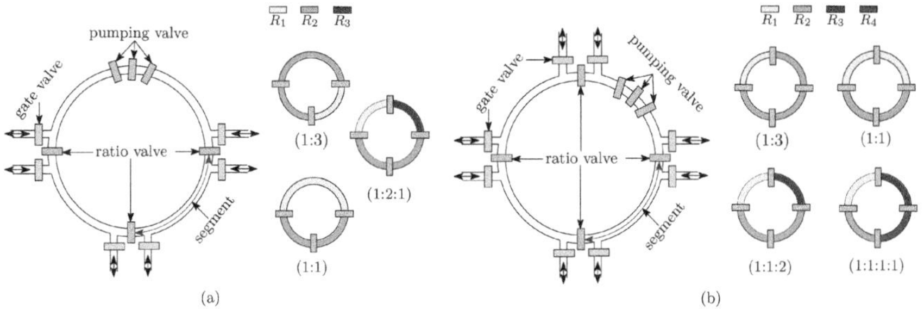

Figure 2.4 Mixing models supported by (a) Ring-3 and (b) Mixer-4.

volume control by using few extra valves compared to Ring-M. Figure 2.4(a) (Figure 2.4(b)) shows the mixing models supported by an unequally (equally) segmented rotary mixer Ring-3 (Mixer-4). Note that in the case of Mixer-N, if each segment of Mixer-N is filled with a mixture of k input reagents $\{R_1 : R_2 : \cdots : R_k = x_1^i : x_2^i : \cdots : x_k^i\}$, for $i = 1, 2, \ldots, N$, the resultant ratio of the mixture will be $\{R_1 : R_2 : \cdots : R_k = \frac{1}{N}\sum_{i=1}^{N} x_1^i : \frac{1}{N}\sum_{i=1}^{N} x_2^i : \cdots : \frac{1}{N}\sum_{i=1}^{N} x_k^i\}$.

The algorithmic aspects of sample preparation with a CFMB have recently been studied by many researchers. Liu *et al.* [LSH15] developed the tree-pruning and grafting algorithm (*TPG*) by transforming an initial mixing graph that is obtained based on the (1:1) mixing model, which is often used in digital microfluidics, e.g., in *twoWayMix* [TUTA08] or *REMIA* [LCH15]. The *TPG* algorithm applies tree-pruning and grafting on the initial dilution graph to transform it to an equivalent one, which is suitable for a special type of unequally segmented rotary mixer (Ring-N). In each mixing step, *TPG* considers only a few specific volumetric ratios supported by $(k : \ell)$ mixing model ($k + \ell = 2^m$, $k, \ell, m \in \mathbb{N}$) available for Ring-$N$. Therefore, *TPG* is unable to exploit the full capability of Ring-N. For example, in Ring-3, the possible mixing models are $(1 : 1), (1 : 3)$, and $(1 : 2 : 1)$; however, *TPG* uses only first two of them.

An improved method called 'volume-oriented sample preparation algorithm' (*VOSPA*) [HLH15] uses a uniformly segmented rotary mixer (Mixer-N) and exploits the power of utilizing multiple intermediate *CF*s. *VOSPA* consists of two phases: master process and subsidiary process. The subsidiary process fills a concentration bank that can be used by the master process, which greedily accumulates various *CF*s from the bank for each segment of the mixer. Note that only one segment of Mixer-N is reused between two calls of the master process, and the remaining $(N-1)$ segments are left unused. Moreover, the subsidiary process can use only two *CF*s in each mixing step for filling the bank that is subsequently used by the master process.

Note that both *TPG* and *VOSPA* produce a mixture of two reagents: sample and buffer. Bhattacharjee *et al.* proposed a satisfiability-based dilution algorithm, namely 'flow-based sample preparation algorithm' (*FloSPA* [BPR$^+$16]), assuming the generalized mixing models supported by an N-segment, continuous-flow, rotary mixer. In the case of dilution, given a target concentration and an error limit, *FloSPA-D* [BPR$^+$16] first minimizes the number of mixing operations, and subsequently, reduces reagent usage. Moreover, two variants of *FloSPA*, namely *FloSPA-M, FloSPA-EM*, are also reported for handling the open problem of k-reagent mixture-preparation ($k \geq 3$) with an N-segment continuous-flow rotary mixer.

During sample preparation, on-chip storage units are needed to store intermediate fluids to be used later. This allows to optimize the reactant costs, to reduce the sample preparation time and/or to achieve the desired ratio. However, the number of storage units is usually limited in given LoC architectures. Since this restriction is not considered by existing methods for sample preparation, the results that are obtained are often found to be useless (in the case when more storage units are required than available) or more expensive than necessary (in the case when storage units are available but not used, e.g., to further reduce the number of mixing operations or reactant cost). We present several storage-aware algorithms for sample preparation with flow-based LoCs addressing the storage constraint issue [BWHB18].

2.5 Summary

In this chapter, a brief review of algorithmic sample preparation with microfluidic biochips is presented. Table 2.2 shows the abbreviated forms of various sample preparation methods. The taxonomy of these algorithms for DMFB and CFMB is depicted in Figure 2.5 and Figure 2.6, respectively, along with the contributions reported in this book (highlighted).

Table 2.2 Sample preparation algorithms and their abbreviations

Single-target dilution algorithms on DMFB	
Algorithm name	Abbreviation
twoWayMix[a] [TUTA08]	Two way mix
DMRW [RBC10]	Dilution and mixing with reduced wastage
IDMA [RBC11]	Improved dilution/mixing algorithm
REMIA [LCH15]	Reactant minimization algorithm
GORMA [CLH13]	Graph-based optimal reactant minimization algorithm
GDA [RBGC14b]	Generalized dilution algorithm
Multiple-target dilution algorithms on DMFB	
PBDA [BBB14]	Pruning-based dilution algorithm
WARA [HLL13]	Waste recycling algorithm
MTC [MRB+14]	Multiple target concentration
NFOSPA [DYHH13]	Network-flow-based optimal sample preparation algorithm
Single-target mixture preparation algorithms on DMFB	
MinMix[b] [TUTA08]	Minimal mixing
RMA [RCK+15b]	Ratio-ed mixing algorithm
MTCS [KRC+13]	Mixing tree with common subtree
CoDOS [LCLH13]	Common dilution operation sharing
GMA [RBGC14b]	Generalized mixing algorithm
Multiple-target mixture preparation algorithms on DMFB	
RSM [HHC14]	Reagent-saving mixing
Ex-CoDOS [LCLH13]	Extended common dilution operation sharing
Droplet streaming algorithms on DMFB	
MTDG [RBGC14a]	Multiple target droplet generation
MDST [RKC+14]	Multiple droplets of single target
Single-target dilution algorithms on CFMB	
TPG [LSH15]	Tree pruning and grafting
VOSPA [HLH15]	Volume-oriented sample preparation algorithm
FloSPA-D [BPR+16]	Flow-based sample preparation algorithm for dilution
Multiple-target dilution algorithms on CFMB	
FloSPA-M [BPR+16]	Flow-based sample preparation algorithm for mixing
FloSPA-EM [BPR+16]	Flow-based sample preparation algorithm for enhanced mixing

[a,b] works also for CFMB.

```
Sample preparation algorithms on DMFB
└─ Dilution
   └─ Single target
      ├─ twoWayMix [TUTA08] (minimum mixing)
      ├─ DMRW [RBC10], IDMA [RBC11] (waste minimization, use
      │  (m:m), m ≥ 1, mixing model)
      ├─ REMIA [LCH15], GORMA [CLH13] (reactant
      │  minimization)
      └─ GDA [RBGC14b] (minimize mixing starting from
         arbitrary input CFs)
   └─ Multiple target
      ├─ PBDA [BBB14] (waste minimization) ............... Chapter 3
      ├─ WARA [HLL13] (reactant minimization)
      ├─ MTC [MRB+14] (minimum on-chip storage)
      └─ NFOSPA [DYHH13] (multiple demand multiple
         target reactant minimization)
   └─ Dilution gradients
      └─ linear [BBH+13b], exponential [BBH+13a], complex
         shaped [BBH+ar] (waste minimization) .......... Chapter 4
└─ Mixture preparation
   └─ Single target
      ├─ MinMix [TUTA08] (minimum mixing)
      ├─ RMA [RCK+15b] (layout aware)
      ├─ MTCS [KRC+13] (waste minimization)
      ├─ CoDOS [LCLH13] (reactant minimization)
      └─ GMA [RBGC14b] (minimize mixing step starting from
         arbitrary ratio of fluids)
   └─ Multiple target
      ├─ RSM [HHC14] (waste minimization)
      └─ Ex-CoDOS [LCLH13] (reactant minimization)
└─ Droplet streaming
   ├─ MTDG [RBGC14a] (waste minimization)
   └─ MDST [RKC+14] (waste minimization)
```

Figure 2.5 Taxonomy of the DMFB sample preparation algorithms.

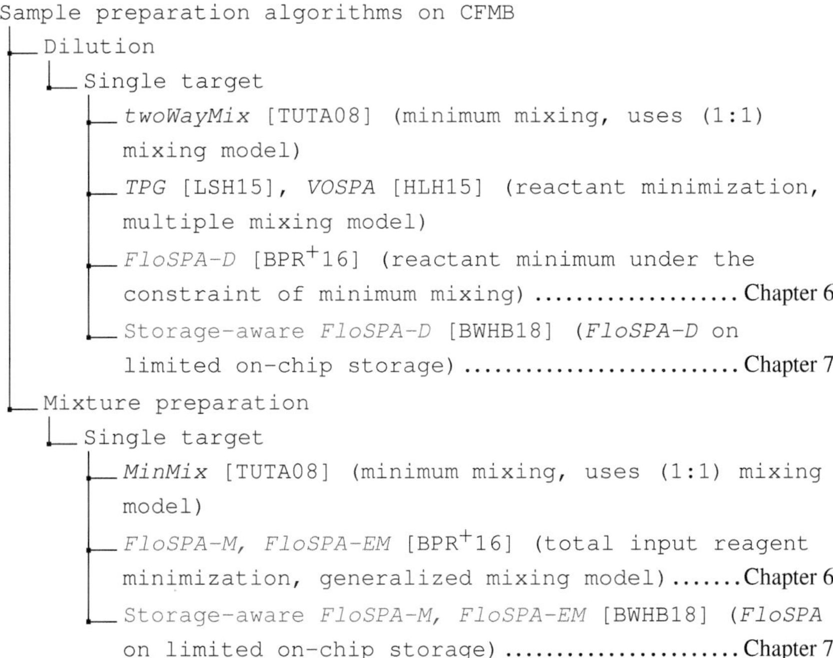

Figure 2.6 Taxonomy of the CFMB sample preparation algorithms.

3

Multiple Dilution Sample Preparation on Digital Microfluidic Biochips

One important step in biochemical sample preparation is dilution, where the objective is to prepare a fluid with a desired concentration factor (CF) by mixing it with buffer solution. The protocols implemented on digital microfluidic biochips (DMFBs) may require several concentration factors of a sample. For example, in Trinders reaction, reference glucose solutions of different concentrations are required [SPF04]. In protein crystallization, for each stock solution, various target concentrations are needed [XCP10]. Bradford protein assay [bio] requires serial dilutions ($10\%, 20\%, 30\%, 40\%,\ 50\%$) of a sample. A linear dilution of concentrations ranging from 10% to 40% sucrose is also required in the case of sucrose gradient analysis [bio]. Multiple target sample preparation can be easily performed with the execution of the dilution steps for each target separately using a single target dilution algorithms [TUTA08, RBC10, LCH15, RBC11, CLH13]. However, this straightforward scheme leads to a large number of mix-split steps and wastage of precious reagents.

In this chapter, we explore a novel strategy of generating multiple target droplets with different concentrations that reduces both the number of mix-split steps and waste droplets. Given a supply of sample and buffer solution, a set of desired target concentration values can be obtained by a sequence of (1:1) mix-split steps, while at each step, an intermediate droplet is allowed to mix with a sample, buffer, or with a previously generated droplet. We report an efficient heuristic to combine some of the waste droplets produced in the process to achieve the target set with a reduced number of mix-split steps and waste production.

3.1 Related Work

Hsieh *et al.* proposed a reagent saving mixing (*RSM* [HHC14]) algorithm for multiple target mixture preparation. In case of multiple target dilution, *RSM* generates each target *CF* using a ratio decomposition step that minimizes the total number of input reagents. After that, *RSM* combines the mixing trees by sharing intermediate droplets. Figure 3.1(a) shows the mixing tree generated by the *RSM* for the target set $\{\frac{11}{64}, \frac{51}{64}\}$. Huang *et al.* proposed a reactant minimization algorithm (*REMIA* [HLC12]) for diluting a sample to the desired concentration factor. In case of multiple target dilution, each individual target *CF*s is generated using a skewed mixing tree. Note that, leaf nodes of an individual mixing tree are prime concentration values (PCV), which can be found by repeatedly mixing sample with buffer droplet. PCVs[1] play key role to sample minimization. For example, in Figure 3.1(b), PCVs for the target *CF*s $\frac{51}{64}$ and $\frac{11}{64}$ are $\{\frac{8}{64}, \frac{16}{64}, \frac{64}{64}\}$ and $\{\frac{4}{64}, \frac{8}{64}, \frac{16}{64}\}$, respectively. After generating each individual target *CF*s, *REMIA* performs a series of exponential dilution operations to create all required nodes of specified PCVs with minimal sample usage. Figure 3.1(b) shows the mixing graph generated for the target set $\{\frac{11}{64}, \frac{51}{64}\}$ using *REMIA*.

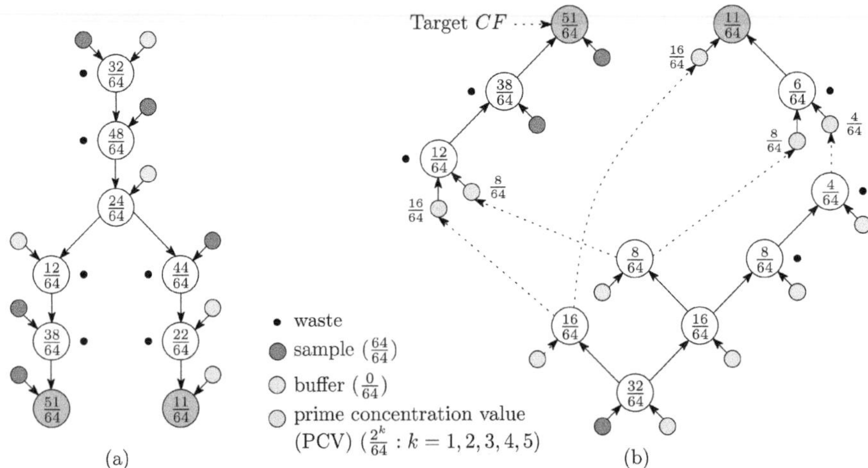

Figure 3.1 Multiple-target dilution using (a) *RSM* [HHC14] and (b) *REMIA* [HLC12].

[1]sample and buffer was also considered as PCV in [HLC12].

3.2 Tree-pruning-based Dilution Algorithm

In sample preparation of two input reagents (dilution) on a DMFB that supports (1:1) mixing model, a mixture of sample and buffer, $\mathcal{D} = \{\langle \text{sample}, c_1\rangle, \langle \text{buffer}, c_2\rangle\}$, where $c_1 + c_2 = 1$ and $0 \leq c_1, c_2 \leq 1$, is approximated as $\{\text{sample} : \text{buffer} = x : y\}$, where $x + y = 2^d, d \in \mathbb{N}$. The value of d is determined by the desired accuracy of approximation (error-tolerance limit, $0 \leq \epsilon < 1$), which is user-specified. The detailed description of the ratio approximation procedure can be found in Section 1.3. In case of dilution, the approximated ratio $\{\text{sample} : \text{buffer} = x : y\}$ can alternatively be represented as the *CF* of sample, i.e., $\frac{x}{x+y}$ or simply by $CF = \frac{x}{2^d}$, where $x + y = 2^d$. Note that, for a fixed value of d, the entire range of *CF*s, which can be generated starting from raw sample ($CF = \frac{2^d}{2^d}$) and buffer ($CF = \frac{0}{2^d}$), are represented as $\frac{1}{2^d}, \frac{2}{2^d}, \ldots, \frac{2^d-1}{2^d}$.

The problem can now be stated as follows: Given a set of n distinct targets $T = \{t_1, t_2, \ldots, t_n\}$, where $t_i \in \left\{\frac{1}{2^d}, \frac{2}{2^d}, \ldots, \frac{2^d-1}{2^d}\right\}$, for $i = 1, 2, \ldots, n$, generate T using only sample and buffer droplets, such that the number of waste droplets is minimized. The number of (1:1) mix-split operations should also be minimized as a secondary objective.

3.2.1 Proposed Methodology

The scheme for generating multiple target concentrations, denoted as pruning-based dilution algorithm (*PBDA*), is described along with an illustrative example. Before discussing the multiple target dilution approach, a few useful definitions and some important observations are presented.

Definition 3.2.1 A dilution tree of order d is defined as a full binary tree [AHU74] with $2^d - 1$ nodes representing the *CF*s $\frac{1}{2^d}, \frac{2}{2^d}, \ldots, \frac{2^d-1}{2^d}$.

In a dilution tree, the node labeled as $\frac{2^{d-1}}{2^d}$ is the root node. The left child of an internal node ($\frac{x}{2^d}$) represents the *CF* generated by mixing the current droplet ($\frac{x}{2^d}$) with buffer ($\frac{0}{2^d}$). Similarly, the right child of $\frac{x}{2^d}$ represents the *CF* generated by mixing the current droplet with sample ($\frac{2^d}{2^d}$). Figure 3.2 shows the dilution tree of order 4. An useful observation for generating a target set from a dilution tree is as follows.

Observation 3.2.1 In order to produce a target set, exploration of only the relevant portion of dilution tree is sufficient.

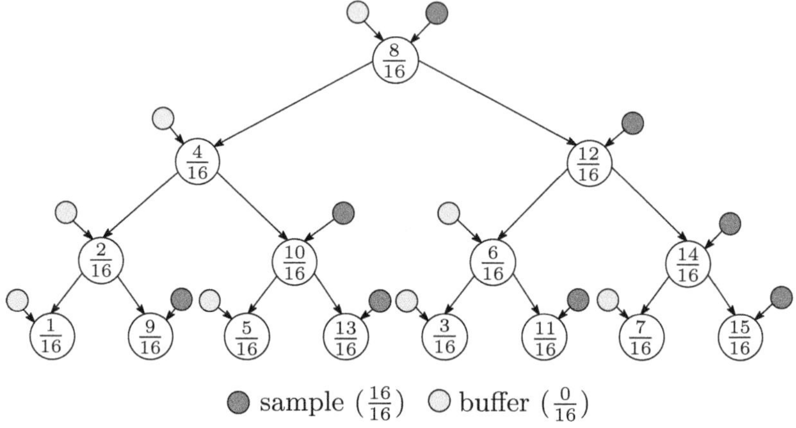

Figure 3.2 Dilution tree of order 4.

In the above observation, "relevant portion" means the minimum subset of nodes in the dilution tree that contains all the targets in a target set. The leaf nodes of the tree represent the targets with two droplets generated therein. Some of the targets may as well be generated in the internal nodes of the dilution tree. If an internal target node has only one child, one need not regenerate that target because one extra droplet is generated in that node. But if an internal target node has two children, then it should be regenerated. A simple example illustrates this observation.

Example 3.2.1 Consider a target set $T = \{\frac{5}{64}, \frac{6}{64}, \frac{8}{64}, \frac{12}{64}, \frac{15}{64}, \frac{20}{64}, \frac{26}{64}, \frac{31}{64}, \frac{34}{64}, \frac{45}{64}, \frac{49}{64}, \frac{57}{64}, \frac{59}{64}, \frac{62}{64}, \frac{63}{64}\}$ to be generated from initial concentrations $\frac{0}{64}$ and $\frac{64}{64}$. The relevant portion of the dilution tree of order 6 is shown in Figure 3.3. One can easily observe from Figure 3.3 that only the target concentrations $\frac{8}{64}$ and $\frac{62}{64}$ need to be regenerated. The total number of (1:1) mix-split steps is found to be 37 along with production of 13 waste droplets. ■

Pruning on the Dilution Tree

We now present another useful observation that motivates us to develop a pruning heuristic on the dilution tree.

Observation 3.2.2 When a target set is generated using the dilution tree, some waste droplets are generated. Moreover, combining two waste droplets may also produce some target droplets.

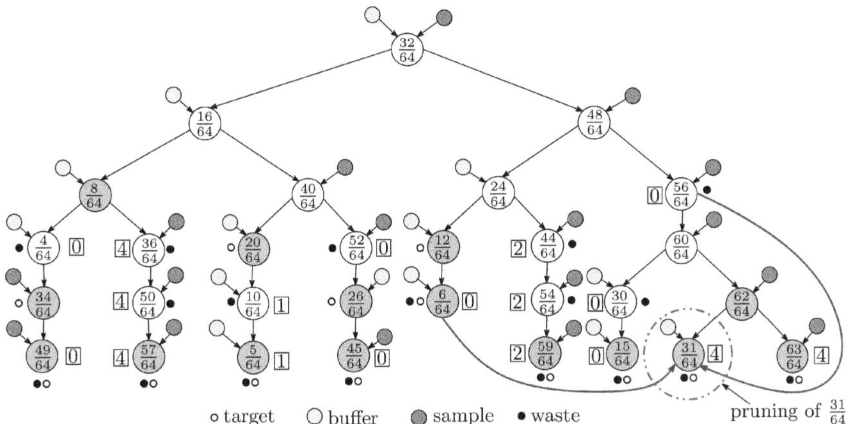

Figure 3.3 Dilution tree for target set T.

There are many waste droplets generated in the dilution tree. Not all waste droplets can be used to generate target droplets. The following example shows how waste droplets can be reused to produce some elements of T in the dilution tree of Figure 3.3.

Example 3.2.2 Consider the dilution tree of Figure 3.3 representing the target set T of Example 3.2.1. The subset of target concentration factors $\left\{\frac{12}{64}, \frac{20}{64}, \frac{26}{64}, \frac{34}{64}\right\}$ should not be used for generating other target concentrations as each target concentration has only one extra droplet generated; however, the intermediate waste droplets with concentrations $\left\{\frac{4}{64}, \frac{10}{64}, \frac{30}{64}, \frac{36}{64}, \frac{44}{64}, \frac{50}{64}, \frac{52}{64}, \frac{54}{64}, \frac{56}{64}\right\}$ can be used for generating other targets. Moreover, for each of the leaf target concentrations $\left\{\frac{5}{64}, \frac{6}{64}, \frac{15}{64}, \frac{31}{64}, \frac{45}{64}, \frac{49}{64}, \frac{57}{64}, \frac{59}{64}, \frac{63}{64}\right\}$ two droplets are generated; one of them can be used for generating other target concentrations. ■

A subset of target droplets can be generated in various ways from waste droplets. For example, the target droplet $\frac{31}{64}$ can be generated by combining waste droplets of $\frac{52}{64}$ and $\frac{10}{64}$. Similarly, waste droplets of $\frac{4}{64}$ and $\frac{6}{64}$ can be used to generate the target $\frac{5}{64}$. One therefore has to decide which target should be generated first and which set of waste droplets is to be used in the process. In order to facilitate this decision, an integer cost is associated with each waste droplet. An intermediate target droplet will have at most one extra droplet, so using it for generating another target is not worthwhile. In the case where a leaf node represents a target, one of them can be used to produce another target droplet. Before defining the cost of waste droplets, we introduce some terminologies.

Definition 3.2.2 The $target\_path$ for a leaf target (t_{leaf}) is defined as the path following the parent of t_{leaf} until another target node or a node with two children is reached.

Definition 3.2.3 A variable $which\_target\_path$ associated with a concentration factor (c) denotes the target concentration whose $target\_path$ contains c.

The following example explains the above definitions.

Example 3.2.3 In Figure 3.3, the $target\_path$ for $\frac{57}{64}$ is $\left(\frac{57}{64}, \frac{50}{64}, \frac{36}{64}\right)$, $target\_path$ for $\frac{5}{64}$ is $\left(\frac{5}{64}, \frac{10}{64}\right)$ and $target\_path$ for $\frac{62}{64}$ is null as it is an internal node. For each *CF* in $\left\{\frac{57}{64}, \frac{50}{64}, \frac{36}{64}\right\}$, $which\_target\_path$ is $\frac{57}{64}$ because each of them belongs to the $target\_path$ of $\frac{57}{64}$. ■

We now define the cost of a leaf target node in the dilution tree of a target set. Informally, the cost of a target leaf node t_{leaf} is defined as the number of mix-split steps that can be saved if t_{leaf} can be generated from other waste droplets. More formally, the cost of a waste droplet is assigned as the cost of $which\_target\_path$ assigned to that waste. Assignment of these costs is performed by Algorithm 1. An example of cost assignment is shown in Figure 3.3 where the number in the square box represents the cost of a waste droplet. The following example illustrates how Algorithm 1 assigns cost to the leaf targets and computes $which\_target\_path$ for the waste droplets in dilution tree of a given target set.

Example 3.2.4 Let us consider the cost assignment to leaf target $\frac{57}{64}$ in Figure 3.3. We start traversing through the parent pointer of $\frac{57}{64}$ until an intermediate target concentration or a node having two children is found. We therefore stop when $\frac{8}{64}$ is reached, which is an intermediate target node. The number of nodes traversed is 2, i.e., $count = 2$. Note that $extra\_cost$ is set to 2 as the number of concentrations from $\frac{8}{64}$ to root $\frac{32}{64}$ is 2. Hence, the cost of $\frac{57}{64}$ is 4. The $target\_path$ of $\frac{57}{64}$ is $\left(\frac{57}{64}, \frac{50}{64}, \frac{36}{64}\right)$ and $which\_target\_path$ for each the concentrations $\left\{\frac{57}{64}, \frac{50}{64}, \frac{36}{64}\right\}$ is $\frac{57}{64}$. ■

Let us interpret the cost of $\frac{57}{64}$. If the concentration $\frac{57}{64}$ is produced from other waste droplets, the right sub-tree of $\frac{8}{64}$ need not be generated. Moreover, the target concentration $\frac{8}{64}$ has one extra droplet. So, $\frac{8}{64}$ need not be regenerated, and we can save four mix-split steps.

After assigning cost to waste droplets, a cost matrix $(M)_{m \times m}$ is defined, where m is the number of waste droplets that can be used from a relevant

Algorithm 1: Cost assignment

Input: Dilution tree *DT* for given a target set T
Output: Cost of waste droplets and $which\_target\_path$ of waste droplets
Initially set the cost of all droplets to 0 and $which\_target\_path$ to dummy concentration having cost 0;
foreach *leaf node* (t_{leaf}) *of* DT **do**
  Set $extra\_cost = 0$;
  Set $count$ = the number of nodes traversed using parent pointer from t_{leaf} until an intermediate target node or an intermediate concentration with two children is found;
  if *an intermediate concentration* (IC) *with two children is found* **then**
    if IC *is target* **then**
      Set $extra\_cost$ = number of nodes from *IC* to root of *DT*;
  Set the *which_target_path* to t_{leaf} for each node along the traversed path from t_{leaf};
  Set the cost of t_{leaf} to $(count + extra\_cost)$;

Algorithm 2: Generate cost matrix

Input: Dilution tree $DT, wastes, m$
// $wastes$ is an m element array of waste droplets.
Output: Cost matrix $(M)_{m\times m}$
Initialize all elements of $(M)_{m\times m}$ by 0;
foreach *(i,j) where* $1 \leq i \leq m, 1 \leq j < i$ **do**
  Let $result = mix\_split(wastes[i], wastes[j])$;
  if $result$ *is a leaf target concentration and* *which_target_path of* $wastes[i]$ *or* $wastes[j]$ *is not equal to* $result$ **then**
    Set $M[i,j]$ = cost of $result$;

portion of the dilution tree. The i-th row of M represents $waste_i$, j-th column of M represents $waste_j$, and $M[i,j]$ represents the number of mix-split step(s) that can be saved by mixing $waste_i$ and $waste_j$ while producing a leaf target droplet. Algorithm 2 shows the essential steps for constructing the cost matrix. Algorithm 2 combines $waste_i$ and $waste_j$ to check whether a leaf target concentration (t_{leaf}) can be produced. If so, it ensures that neither of $waste_i$ or $waste_j$ arrives from $target\_path$ of t_{leaf}.

Based on Algorithm 1 and Algorithm 2, we can prune the dilution tree. The pruning strategy is described in Algorithm 3. It first assigns the cost and sets $which\_target\_path$ for each waste droplet. Next, the cost matrix M is constructed. This is followed by choosing, in a greedy manner, the largest

Algorithm 3: Pruning

Input: Dilution tree DT
Output: Pruned dilution tree.
Assign cost to each waste droplets. (Algorithm 1);
Create cost matrix (M). (Algorithm 2);
while *true* **do**
  Find the maximum element in $M[i, j]$ where $cost(wastes[i]) + cost(wastes[j])$ is minimum;
  // $cost(c)$ returns the cost of concentration c
  Assume that the maximum occurs at (i, j)-th index of M;
  if $M[i, j] \leq 1$ **then**
    return;
  Set $target_i$ to *which_target_path* of $wastes[i]$;
  Set $target_j$ to *which_target_path* of $wastes[j]$;
  Set cost of $target_i$ and $target_j$ to 0;
  Prune the sub-path containing the removed target as leaf node from DT;
  Remove $wastes[i]$, $wastes[j]$ and the sub-path that contains the target from $wastes$;
  Construct cost matrix (M) (Algorithm 2) for modified waste droplet set;

cost target droplet that can be pruned. It may be noted that the maximum may occur multiple times in M for different (i, j) pairs. Algorithm 3 chooses the pair (i, j) where the cumulative cost of $waste_i$ and $waste_j$ is minimized. Before pruning, it sets the cost of $which\_target\_path$ for $waste_i$ and $waste_j$ to 0. It prevents leaf targets from pruning that have $waste_i$ or $waste_j$ in their $target\_path$. It may so happen that $waste_i$ or $waste_j$ does not belong to any $target\_path$ of a leaf target. In this case, setting its cost to 0 has no effect as it has only one waste droplet and it is used in pruning. After pruning, the set of waste droplets is modified and the cost matrix is recalculated for further pruning. The following example explains the steps of Algorithm 3.

Example 3.2.5 The initial cost matrix for the waste droplet set $\left\{\frac{4}{64}, \frac{5}{64}, \frac{6}{64}, \frac{10}{64}, \frac{15}{64}, \frac{30}{64}, \frac{31}{64}, \frac{36}{64}, \frac{44}{64}, \frac{45}{64}, \frac{49}{64}, \frac{50}{64}, \frac{52}{64}, \frac{54}{64}, \frac{56}{64}, \frac{57}{64}, \frac{59}{64}, \frac{63}{64}\right\}$ has a maximum entry of 4. There are several valid waste droplet combinations, which generate targets of cost 4. Table 3.1 shows the different alternatives of cost 4 targets. The highlighted row shows the minimum cumulative cost of two waste droplets that produce $\frac{31}{64}$ of cost 4. The dilution tree after pruning $\frac{31}{64}$ is shown in Figure 3.4. The concentrations $\frac{56}{64}$ and $\frac{6}{64}$ have no waste droplet, so they are removed from further consideration along with $\frac{31}{64}$. Now the new waste set becomes $\left\{\frac{4}{64}, \frac{5}{64}, \frac{10}{64}, \frac{15}{64}, \frac{30}{64}, \frac{36}{64}, \frac{44}{64}, \frac{45}{64}, \frac{49}{64}, \frac{50}{64}, \frac{52}{64}, \frac{54}{64}, \frac{57}{64}, \frac{59}{64}, \frac{63}{64}\right\}$ and the

Table 3.1 Different alternatives for maximum entry (4) in the initial cost matrix

$waste_1$	$waste_2$	cost($waste_1$)	cost($waste_2$)	Total Cost	Target	cost($target$)
$\frac{5}{64}$	$\frac{57}{64}$	1	4	5	$\frac{31}{64}$	4
$\frac{52}{64}$	$\frac{10}{64}$	0	1	1	$\frac{31}{64}$	4
$\frac{57}{64}$	$\frac{6}{64}$	4	0	4	$\frac{31}{64}$	4
$\frac{56}{64}$	$\frac{6}{64}$	0	0	0	$\frac{31}{64}$	4
$\frac{59}{64}$	$\frac{4}{64}$	2	0	2	$\frac{31}{64}$	4
$\frac{56}{64}$	$\frac{59}{64}$	0	2	2	$\frac{57}{64}$	4
$\frac{63}{64}$	$\frac{52}{64}$	4	0	4	$\frac{57}{64}$	4

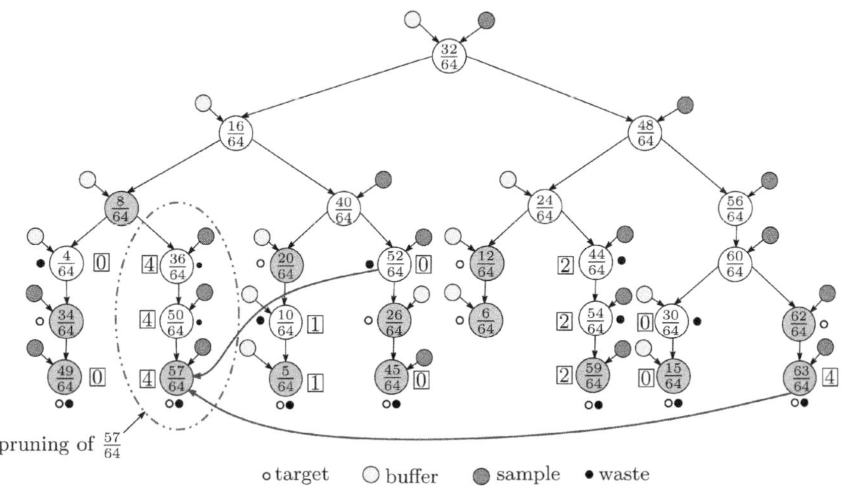

Figure 3.4 Tree after pruning $\frac{31}{64}$.

cost matrix is recalculated. The new cost matrix has only one entry of cost 4. Hence, $\frac{52}{64}$ and $\frac{63}{64}$ are combined and target concentration $\frac{57}{64}$ is generated. The dilution tree after pruning $\frac{57}{64}$ is shown in Figure 3.5. The concentrations $\frac{52}{64}, \frac{63}{64}, \frac{57}{64}, \frac{50}{64}, \frac{36}{64}$ are removed and the modified waste droplet set becomes $\left\{\frac{4}{64}, \frac{5}{64}, \frac{10}{64}, \frac{15}{64}, \frac{30}{64}, \frac{44}{64}, \frac{45}{64}, \frac{49}{64}, \frac{54}{64}, \frac{59}{64}\right\}$. The new cost matrix has all entries 0 and thus no further pruning is possible. Figure 3.5 shows the final pruned tree. One can easily verify that after pruning, all target droplets have been generated. However, it may so happen that after pruning, there are still some targets left. In such a case, the dilution tree for the remaining target droplets has to be generated separately and this process continues until all targets are produced. ■

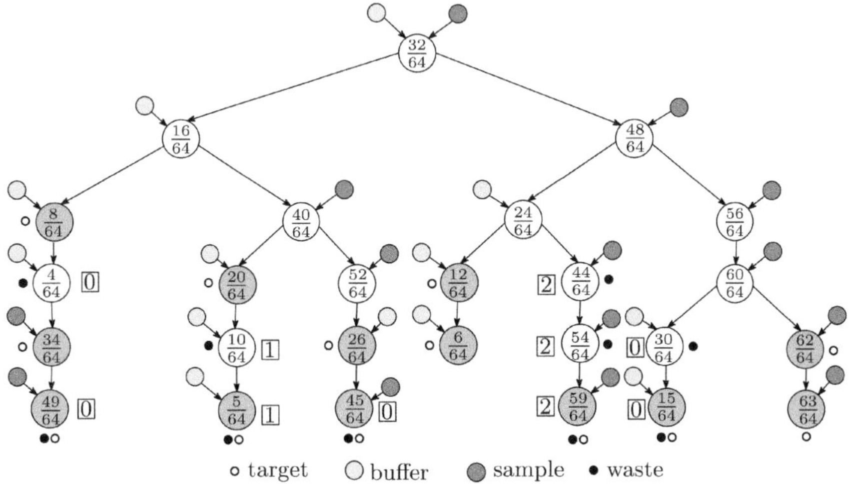

Figure 3.5 Final pruned tree.

In the above example, the total number of (1:1) mix-split steps required is 26, and 5 waste droplets are generated, i.e., we obtain 30% savings in mix-split steps and 62% savings for waste droplets over the base approach when pruning is applied on the dilution tree. The running time of the pruning algorithm is quadratic in the number of waste droplets. One need not explicitly calculate the cost matrix every time if the entries for waste droplets are sorted initially. For each leaf target sorted in the decreasing order of cost, one can find all valid combinations of waste droplets that generate a particular leaf target in $O(n_w)$ time and choose the one with least cumulative cost; here, n_w denotes the number of waste droplets that can be considered for generating a particular leaf target in the dilution tree. Note that pruning can occur only for leaf target nodes. For a dilution tree with T_{leaf} leaf targets, each pruning step therefore requires $O(T_{leaf} \times n_w)$ operations in the worst case.

3.3 Experimental Results

To demonstrate the effectiveness of the pruning-based dilution algorithm (*PBDA*), it is compared with those reported by *RSM* [HHC14] and *REMIA* [HLC12]. A prototype tool is implemented to perform simulation in Python. For simulation purposes, *d* is set to 10. A total of 5000 random target sets of different sizes (10, 30, 50, 100) are considered for simulation purpose. For each of the 5000 target sets, %-savings in the number of mix-split steps

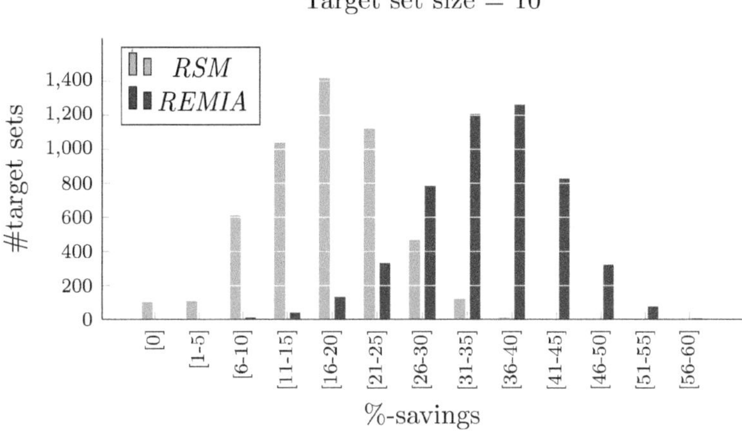

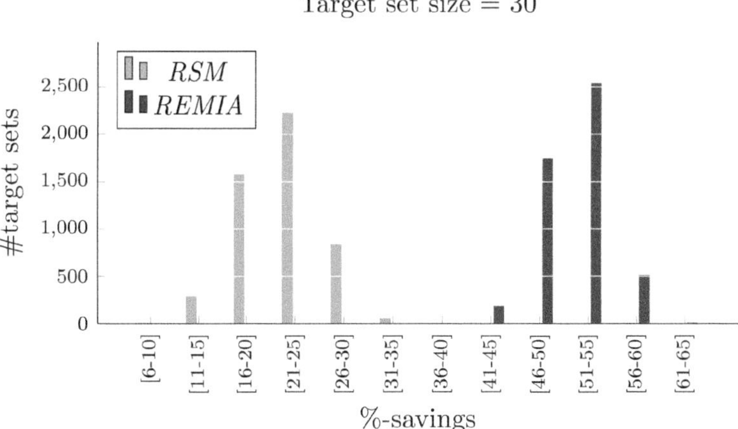

Figure 3.6 Histogram of %-savings in the mix-split steps of *PBDA* compared to *RSM* [HHC14] and *REMIA* [HLC12].

and in the waste count of the *PBDA* are calculated with respect to two earlier methods. The histogram plots of %-savings are shown in Figure 3.6 and Figure 3.7. In these plots, the horizontal axis denotes the %-savings and the vertical axis shows the number of target sets with a particular value of %-savings. The following expression is used for computing the %-savings.

$$\text{mix-split}_{\text{savings}} = \frac{\text{mix-split}_{\text{other}} - \text{mix-split}_{PBDA}}{\text{mix-split}_{\text{other}}} \times 100\% \qquad (3.1)$$

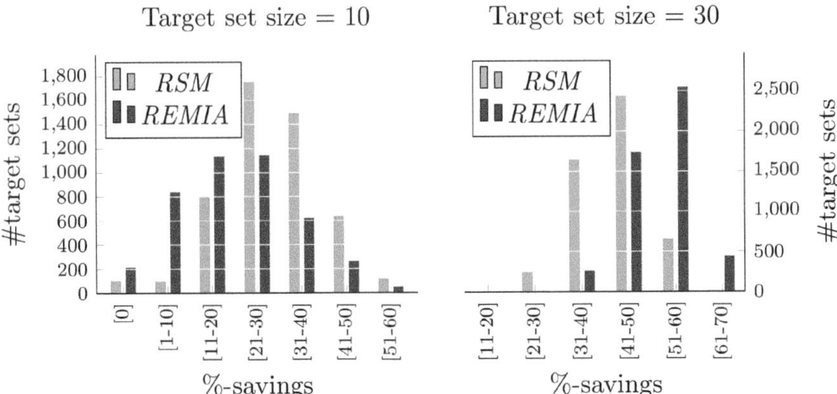

Figure 3.7 Histogram of %-savings in waste droplets of *PBDA* compared to *RSM* [HHC14] and *REMIA* [HLC12].

$$\text{waste}_{\text{savings}} = \frac{\text{waste}_{\text{other}} - \text{waste}_{PBDA}}{\text{waste}_{\text{other}}} \times 100\% \tag{3.2}$$

Experimental results reveal that with increasing cardinality of the target set, *PBDA* scales up favorably. This happens because of the fact that the effect of pruning becomes more dominant when the target set size increases. As a result, improved solutions are obtained that yield fewer mix-split steps and reduced wastage. Target sets of size 10, 30, 50, and 100 are considered in the experiments. Figure 3.8 shows the average behavior of the *PBDA* against *RSM* [HHC14] and *REMIA* [HLC12]. The results demonstrate significant savings in terms of both mix-split steps and waste droplets. Additionally, sample and buffer consumption of *PBDA* is compared with those obtained

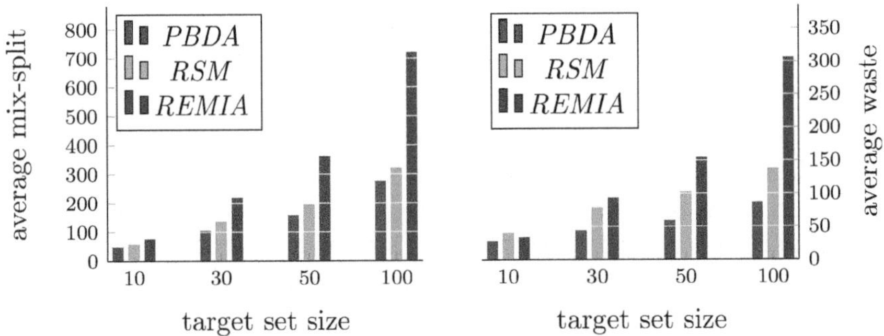

Figure 3.8 The average number of mix-split steps and waste droplets for *PBDA*, *RSM* [HHC14], and *REMIA* [HLC12].

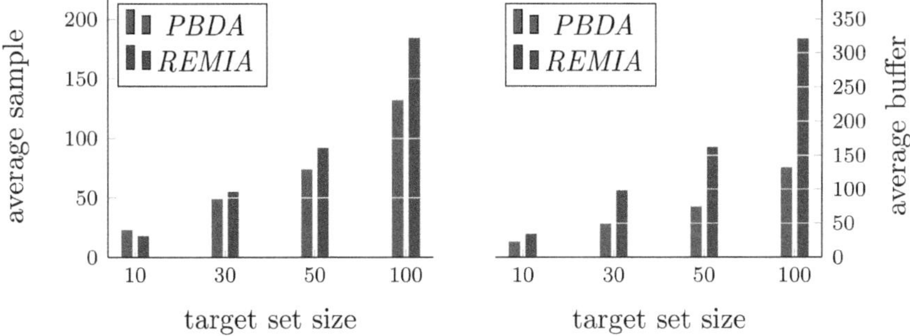

Figure 3.9 Consumption of the average sample and buffer droplets for *PBDA* and *REMIA* [HLC12].

by a reactant-minimizing algorithm *REMIA* [HLC12]. Figure 3.9 shows that *PBDA* requires fewer sample droplets for a target set of size 30 or more, as well as fewer buffer droplets. Note that, in *RSM* and *REMIA*, the possibility of mixing two intermediate droplets, i.e., pruning, was not explored. The performance of *PBDA* is found to be significantly improved when pruning is applied.

In *PBDA*, a target droplet can be generated in two ways: either by combining the current droplet with a sample/buffer droplet or by mixing two previously generated droplets, i.e., by applying the aforesaid pruning procedure. Thus, the number of pruning has a direct consequence on the storage requirement in the realization of *PBDA*. Before pruning, a droplet generated earlier is to be stored until it is mixed with a desired droplet. The extra storage required for pruning is bounded above by the number of maximum pruning operations that can occur in the dilution tree. In order to estimate the additional storage cells experimentally, an extensive simulation on various examples is performed. Since some applications may require a large number of target concentrations (up to 100) [XCP10], 5000 synthetic random target sets of size 10, 30, 50, and 100 are simulated. From Figure 3.10, it can be observed that the maximum pruning count is 21; thus, for implementing the test data, at most 21 extra storage cells are required. However, for a target set size up to 50, only 15 extra storage cells will be sufficient. The pruning algorithm can also be adapted based on the available storage cells. In this case, the number of pruning should be restricted in order to satisfy the limited availability of storage cells. A separate stack of d storage cells is required for generating the targets, which are present on the final dilution tree. Here, a stack is used so that the dilution tree can be processed in depth-first order [Tar72].

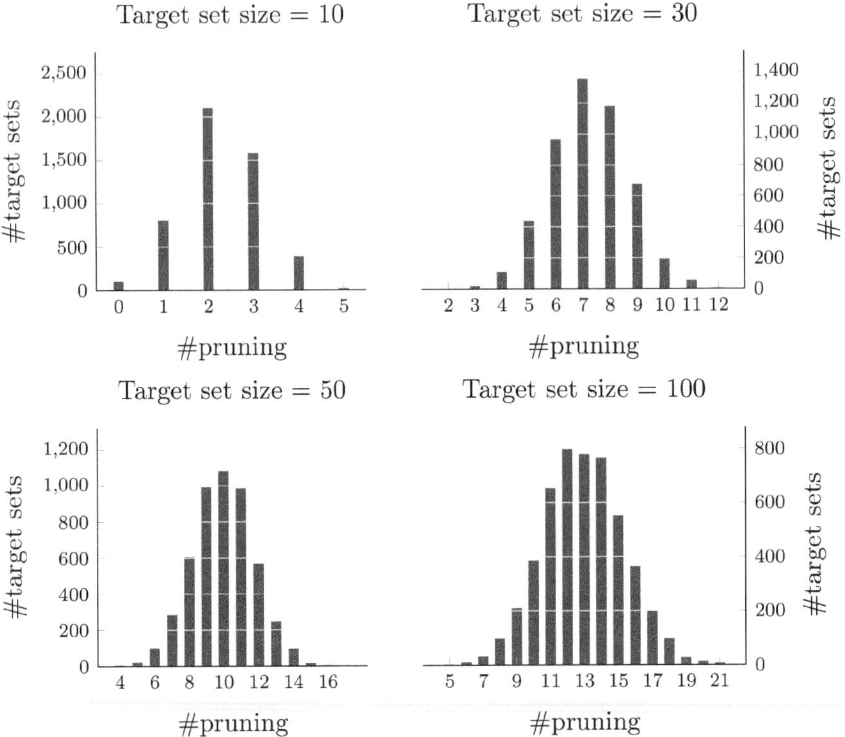

Figure 3.10 Histogram of pruning for random target sets of different size.

3.4 Conclusions

This this chapter, a multiple target dilution algorithm (*PBDA*) is presented. In *PBDA*, the notion of dilution tree has been introduced for representing the entire range of target concentration factors. *PBDA* constructs only the relevant portion of the dilution tree and expedites the search, based on pruning heuristic. Simulation results reflect the improvement of the *PBDA* in comparison with other approaches (*RSM* and *REMIA*). The efficiency of *PBDA* increases in terms of mix-split (sample preparation time) and waste (sample cost) with the increase of target set size. Moreover, it also saves valuable reagents (sample and buffer).

4

Efficient Generation of Dilution Gradients with Digital Microfluidic Biochips

Digital microfluidic biochips (DMFBs) are now being extensively used to automate several biochemical laboratory protocols such as clinical analysis, point-of-care diagnostics, or DNA sequencing. In many biological assays, e.g., bacterial susceptibility tests and cellular response analysis, samples or reagents are required in multiple concentration (or dilution) factors, satisfying certain gradient patterns such as linear, exponential, or parabolic. Dilution gradients play an essential role in *in vitro* analysis of many biochemical phenomena, including growth of pathogens and selection of drug concentration. For example, in drug design, it is important to determine the minimum amount of an antibiotic that inhibits the visible growth of bacteria isolate (defined as *minimum inhibitory concentration* (MIC) [And01]). The drug with the least concentration factor (i.e., with the highest dilution) that is capable of arresting the growth of bacteria is considered as MIC. During the past decade, a variety of automated systems for bacterial identification and antimicrobial susceptibility test have been developed [NGL+12, CHDW12, DVGL94, CWJ12], which provide results in only few hours rather than days, compared to traditional overnight procedures. Dilution gradients are also required for *in vitro* simulation of cellular analysis that are deployed to replicate *in vivo* mechanisms of human-physiological functions [WLP17]. In living organisms, cells are surrounded by various fluids with concentration gradients, which influence growth, inflammation, cancer metastasis, wound healing, and the effectiveness of drug delivery to cells [SHK10, KF08, GHW+18]. In order to observe, *in vitro*, cellular responses to molecular gradients, the generation of concentration profiles is needed [WC10, LKK+11, SWLJ06, RWX+09]. Additionally, molecular-concentration gradients are required for *in vitro* toxicity analysis [SIP16b, ZWQ12], studies in enzymatic kinetics [SI03], and protein expressions [COW+08].

Dilution gradients are traditionally prepared using continuous-flow microfluidic devices. Unfortunately, most of them suffer from inflexibility and non-programmability, and they require large volumes of costly stock solutions. DMFBs, on the other hand, are shown to produce, more efficiently, samples with multiple dilution factors. However, none of the existing DMFB-based algorithms [HLC12, HLL13, MRB$^+$14, HHC14] utilize the properties of the gradient-profile while optimizing reactant cost and sample-preparation time. In this chapter, we explore the underlying combinatorial attributes of different gradients and harnessed them for efficient production of the desired concentration profile. For linear gradients, we present theoretical results concerning the number of mix-split operations and waste production and prove an upper bound on on-chip storage requirement. A cost-effective method for generating a wide class of exponential gradients is also discussed. Finally, in order to handle a complex-shaped gradient, we posit a digital-geometric technique to approximate it with a sequence of linear gradients.

4.1 Literature Review

Dilution gradients are traditionally prepared by using continuous-flow microfluidic ladder networks [WJWL10], or by other networks of microchannels [DCJW01, SWJ08, JHK$^+$11, LKA$^+$09, GSO$^+$15, WSR$^+$13] based on convective or diffusive mixing of two or more streams [LCWF11a]. The degree of diffusion can be regulated by the flow rate, channel geometry, and channel dimension. Designs of such gradient generators on a continuous flow microfluidic biochip (CFMB) were reported in [WMRRN10, OMRW06, FPM12, WBG16, SBV11]. The flow rates were adjusted by controlling the channel length, which is proportional to fluidic resistance in each channel, or by changing inlet cross sections or injection velocities [BGB17]. Several designs of 2D combinatorial dilution-gradient generator have also been reported [JHK$^+$11, TGZ$^+$17]. A comprehensive review of flow-based microfluidic gradient-generation systems can be found in [WLP17, SIP16a]. A promising valve array-based fully programmable microfluidic platform was proposed in [FM11] offering more flexibility and reconfigurability in bioassay operations. A general-purpose sample-preparation algorithm was proposed to implement dilution and mixing in the same framework on a CFMB that supports multiple mixing models [BPR$^+$16]. However, no algorithmic approach for generating dilution gradients was proposed so far on such programmable microfluidic platform. 2D array-based continuous-flow microfluidic structures have also been proposed for producing certain concentration profiles [WBG16]. An inexpensive thread-based gradient generator is proposed in [RRC+16] that uses high-frequency sound waves to control convective transport through the thread network.

4.2 Linear Gradient

Among various types of dilution profiles, linear gradient is most widely used for biochemical analysis [WCX+15, BKE16]. A sequence of *CF*s corresponds to a *linear gradient* if they appear in arithmetic progression [BBH+13b]. Motivated by an example described by Brassard *et al.* [BMMG+11], we present an algorithm for producing any given linear dilution gradient with minimum wastage. In other words, total reagent consumption is minimized. We assume that two boundary-*CF*s (first and last *CF*s) of the target sequence are available to the gradient-generator. If droplets with two boundary-*CF*s are not supplied, one can prepare them by diluting the original sample (100%) with buffer (0%) following some earlier algorithm [HLC12, TUTA08, RBGC14a]. Design of the gradient generator is based on two observations, which are self-evident.

1. Let a sample be diluted with the same buffer to produce two *CF*s denoted by C_1 and C_2. If a droplet of $CF = C_1$ is mixed with a droplet of $CF = C_2$ in (1:1) mixing ratio, the *CF* of the resulting mixture will be $\frac{C_1+C_2}{2}$, i.e., the *mean value* of C_1 and C_2.
2. Let C_1 and C_2 represent *CF*s appearing in a linear gradient. If C_1 and C_2 are separated by an odd number of elements, then (1:1) mixing of C_1 and C_2 will produce a fluid with $CF = C_3$, where C_3 is the *median* of the subsequence of *CF*s bounded within C_1 and C_2.

   Example 4.2.1 Let 15%, 20%, 25%, 30%, 35%, 40%, 45% represent a linear gradient of *CF*s. If we mix two droplets with *CF*s 20% and 40% in (1:1) mixing ratio, the resulting *CF* will be 30%, which is the median of the sub-sequence 20%, 25%, 30%, 35%, 40%. ■

The above observations are used to design an algorithm for producing a given linear gradient with no wastage. Moreover, only those concentrations that are elements of the gradient set will be generated during this process.

Let $\mathcal{L} = \{\frac{a}{2^d}, \frac{a+n}{2^d}, \frac{a+2n}{2^d}, \frac{a+3n}{2^d}, \cdots, \frac{a+2^k n}{2^d}\}$, $|\mathcal{L}| = 2^k + 1$, where $k, n, d \in \mathbb{N}$ is a linear gradient of targets to be generated from raw sample

($\frac{2^d}{2^d}$) and buffer ($\frac{0}{2^d}$). However, when $|\mathcal{L}| \neq 2^k + 1$, one can split $\mathcal{L}$ into few smaller linear gradients and generate them separately. The process of generating the target *CF*s satisfying a linear dilution gradient can be envisaged as a tree structure called *linear dilution tree* described below. We define $\mathcal{L}' = \mathcal{L} \setminus \{\frac{a}{2^d}, \frac{a+2^k n}{2^d}\} = \{\frac{a+n}{2^d}, \frac{a+2n}{2^d}, \cdots, \frac{a+(2^k-1)n}{2^d}\}$. We will show shortly that $\mathcal{L}'$ can be generated with zero waste.

Definition 4.2.1 A linear dilution tree (LDT) is a complete binary search tree having $2^k - 1$ nodes, where each node represents a *CF*-value in the target set $\mathcal{L}'$, where $|\mathcal{L}'| = 2^k - 1$. The nodes labeled with $CF = \frac{a+n}{2^d}, \frac{a+3n}{2^d}, \frac{a+5n}{2^d}, \cdots,$ $\frac{a+(2^k-1)n}{2^d}$ appear as leaf nodes of LDT, and the median-*CF* of $\mathcal{L}'$ appears as the root of LDT. Also, the *CF* represented by any non-leaf node of LDT will be the mean value of the *CF*s corresponding to its two children.

Thus, LDT will have a depth of $(k - 1)$, where the root is assumed to be at depth 0. Algorithm 1 builds an LDT from the input target set, on which Algorithm 2 is run to produce the droplets in the target set $\mathcal{L}'$.

Let us consider an example to illustrate the gradient generation method.

Example 4.2.2 Let $\mathcal{L} = \{\frac{a}{2^d}, \frac{a+n}{2^d}, \frac{a+2n}{2^d}, \ldots, \frac{a+8n}{2^d}\}$ be a linear gradient of targets to be generated from $\frac{0}{2^d}$ and $\frac{2^d}{2^d}$, i.e., $|\mathcal{L}| = 2^3 + 1 = 9$. The corresponding linear dilution tree (LDT) is shown in Figure 4.1. LDT will be

---

Algorithm 1: LDT

Input: A set of CFs $\mathcal{L}$, $|\mathcal{L}| = 2^k + 1$
Output: Linear dilution tree with $|\mathcal{L}'|$ nodes, $\mathcal{L}' = \mathcal{L} \setminus \{\frac{a}{2^d}, \frac{a+2^k n}{2^d}\}$

```
if L' contains only one CF then
    Create a leaf v storing this CF;
else
    Let C_mid be the median of L';
    Set L_left = CFs less than C_mid in L';
    Set L_right = CFs greater than C_mid in L';
    T_left = LDT(L_left);
    T_right = LDT(L_right);
    Create a node v with CF = C_mid;
    Make T_left the left sub-tree of v;
    Make T_right the right sub-tree of v;
return v;
```

Algorithm 2: Linear dilution-gradient

Input: $\mathcal{L} = \left\{\frac{a}{2^d}, \frac{a+n}{2^d}, \frac{a+2n}{2^d}, \ldots, \frac{a+2^k n}{2^d}\right\}$
Output: Output ordering of CFs in $\mathcal{L}$
$\mathcal{L}' = \left\{\frac{a+n}{2^d}, \frac{a+2n}{2^d}, \ldots, \frac{a+(2^k-1)n}{2^d}\right\}$;
Concentration factors $\frac{a}{2^d}$ and $\frac{a+2^k n}{2^d}$ are supplied;
return postorder(LDT($\mathcal{L}'$));
// postorder(T) is the post-order traversal of the binary tree T

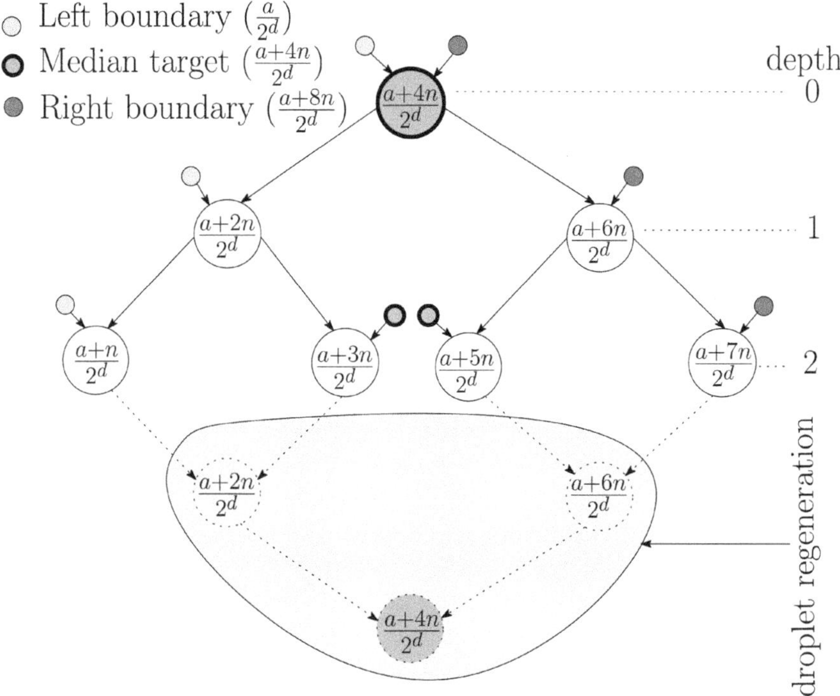

Figure 4.1 Linear dilution tree.

traversed in the depth-first order to produce the target droplets in a post-order mixing sequence. Let us assume that the two boundary-*CF*s $\frac{a}{2^d}$ and $\frac{a+8n}{2^d}$ are generated from the sample ($\frac{2^d}{2^d}$) and buffer ($\frac{0}{2^d}$) by using any conventional dilution algorithm [HLC12, TUTA08, RBGC14a]. Initially, two droplets are generated with $CF = \frac{a+4n}{2^d}$ by mixing one droplet of $\frac{a}{2^d}$ and $\frac{a+8n}{2^d}$ each (represented as the root in Figure 4.1). One of these droplets is stored and the other one is mixed with a droplet of $CF = \frac{a}{2^d}$ to produce two droplets of

$CF = \frac{a+2n}{2^d}$. Again, one of them is stored and the other one is mixed with a droplet of $CF = \frac{a}{2^d}$ to generate two droplets of $CF = \frac{a+n}{2^d}$ (leftmost leaf), out of which one droplet is sent to the output and the other one is stored. Next, the two droplets with *CF*s $\frac{a+2n}{2^d}$ and $\frac{a+4n}{2^d}$, which were stored in the first two steps, are mixed to produce two droplets of $CF = \frac{a+3n}{2^d}$. One of them is sent to the output; the remaining one is mixed with the one with $CF = \frac{a+n}{2^d}$ stored in the third step. This step *regenerates* the two droplets with $CF = \frac{a+2n}{2^d}$, both of which were consumed in earlier steps (shown as dotted circle). One of them is stored again and now the other one is transported to the output. Similar (1:1) mix-split sequences are performed on the right-half of LDT in post-order fashion, and finally, two droplets of $CF = \frac{a+4n}{2^d}$ are regenerated by mixing droplets with $CF = \frac{a+6n}{2^d}$ and $\frac{a+2n}{2^d}$. It may be observed that "no waste droplet" is produced for generating the entire linear dilution sequence $\mathcal{L}'$. Only one droplet for every non-boundary-*CF*-value in the gradient is produced, excepting the median-*CF* (i.e., the root), for which two droplets are produced. ■

The following observations are now immediate.

Observation 4.2.1 The droplets with boundary-*CF*s are used only along the leftmost and the rightmost root-to-leaf path in LDT.

Observation 4.2.2 *(Regeneration principle)* All droplets with the *CF*-values corresponding to each internal node of LDT are used in subsequent mixing operations after their production, and are regenerated later for replenishment.

The *regeneration principle* that is utilized here is a unique feature of the linear gradient generation method. In all multi-target generation algorithms [HHC14, HLC12, HLL13], for each target *CF*, at least one target droplet is kept aside (for final output) once it is produced.

The following two lemmas lead to Theorem 4.2.3, which will be used later to calculate the number of mix-split steps, waste droplets, and the number of boundary droplets that are required to produce a linear gradient of size $2^k + 1$.

Lemma 4.2.1 During the process of gradient generation, the *CF* corresponding to each node at depth i is produced as $2^{k-i}+2$ droplets (for $i < k$), and as two droplets when $i = k$ (leaf node), when the size of linear dilution gradient ($\mathcal{L}$) is $2^{k+1} + 1$.

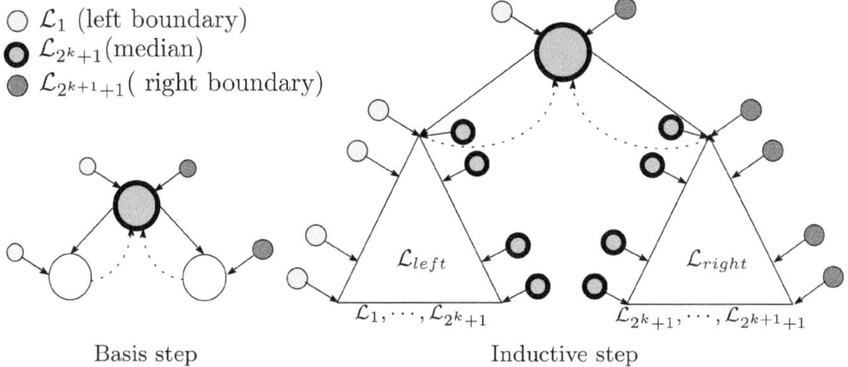

Figure 4.2 Recursive structure of the linear dilution tree.

Proof. We prove the lemma using induction on k, given the target-set size $|\mathcal{L}| = 2^{k+1} + 1$.

Basis: For $k = 1$, $|\mathcal{L}| = 2^{1+1} + 1 = 5$; in this case, we need to generate three *CF*-values from two boundary-*CF*s. Figure 4.2 shows the linear dilution tree. The total number of droplets corresponding to the median target-*CF* (at the root, depth $= 0$) is $4(= 2 + 2)$, and the target concentration at depth 1 is 2 each. Hence, Lemma 4.2.1 is true for $k = 1$.

Induction hypothesis: Assume the statement is true for all $m \leq k - 1$.

Inductive steps: Consider the target set $\mathcal{L}$ of size $2^{k+1} + 1$, i.e., $m = k$. One can split $\mathcal{L}$ into three parts: $\mathcal{L}_{\text{left}}$ that contains the first $2^k + 1$ targets of $\mathcal{L}$, i.e., $\mathcal{L}_{\text{left}} = \{\mathcal{L}_1, \ldots, \mathcal{L}_{2^k+1}\}$; $\mathcal{L}_{\text{median}} = \{\mathcal{L}_{2^k+1}\}$; and $\mathcal{L}_{\text{right}} = \{\mathcal{L}_{2^k+1}, \ldots, \mathcal{L}_{2^{k+1}+1}\}$. The elements in $\mathcal{L}_{\text{left}}$ can be generated by using $\mathcal{L}_1$ and $\mathcal{L}_{2^k+1}$ as boundary targets. Similarly, those in $\mathcal{L}_{\text{right}}$ can be generated by using $\mathcal{L}_{2^k+1}$ and $\mathcal{L}_{2^{k+1}+1}$ as boundary targets. One can easily generate $\mathcal{L}$ by using $\mathcal{L}_1$ and $\mathcal{L}_{2^{k+1}+1}$ as boundary targets (see Figure 4.2). By induction hypothesis, the number of each droplet generated during the process at depth i of $\mathcal{L}_{\text{left}}$ and $\mathcal{L}_{\text{right}}$ is $2^{k-1-i} + 2$ when $i < k - 1$, and is 2 when $i = k - 1$. Ignoring the regeneration part, the number of each droplet generated during the process at depth i of $\mathcal{L}_{\text{left}}$ and $\mathcal{L}_{\text{right}}$ is 2^{k-1-i} when $i < k - 1$, and is 2 at depth $k - 1$. From Observation 4.2.1, it follows that $\mathcal{L}_{\text{median}}$ is used only in the rightmost path of $\mathcal{L}_{\text{left}}$ and in the leftmost path of $\mathcal{L}_{\text{right}}$ as shown in Figure 4.2. By inductive hypothesis, the total number of droplets generated (ignoring the regeneration part) is $2 \times (\sum_{i=0}^{k-2} 2^{k-1-i} + 2) = 4 \times (\sum_{i=0}^{k-2} 2^i + 1) = 4 \times (2^{k-1} - 1 + 1) = 4 \times 2^{k-1}$. Hence, the required number of $\mathcal{L}_{\text{median}}$

droplets is $\frac{4\times 2^{k-1}}{2} = 2^k$. Since the number of regenerated droplets is 2, the total number of droplets generated at the root (depth = 0) is $2^k + 2$. This completes the proof. □

Lemma 4.2.2 The number of droplets required for each boundary-*CF* is 2^k for $k \geq 1$, where the target linear dilution gradient ($\mathcal{L}$) size is $2^{k+1} + 1$.

Proof. From Observation 4.2.1, it follows that boundary droplets are needed only for the nodes lying on the leftmost and rightmost paths of LDT. Note that the regeneration process for an internal node does not require any boundary droplet. So, the number of droplets generated excluding regeneration is 2^{k-i} at depth i and 2 at depth k, along the left-most or rightmost path in LDT. The number of droplets along these paths is $\sum_{i=0}^{k-1} 2^{k-i} + 2$. Hence, the total number of required boundary droplets will be given by:

$$\frac{\sum_{i=0}^{k-1} 2^{k-i} + 2}{2} = \left(\sum_{i=0}^{k-1} 2^i\right) + 1 = (2^k - 1) + 1 = 2^k \qquad \square$$

These boundary-*CF*s can be efficiently produced from the raw sample (100%) and buffer (0%) by running a dilution engine [RBGC14a].

Theorem 4.2.3 Algorithm 2 generates a linear dilution gradient $\mathcal{L}$ of size $2^{k+1} + 1$ in $2^{k-1}(k + 4) - 1$ mix-split steps without producing any waste droplets, when 2^k droplets of each boundary-*CF* are supplied.

Proof. The LDT has $2^{k+1} - 1$ nodes including 2^k leaf nodes. Each leaf node requires only one mix-split operation. By Lemma 4.2.1, the number of each droplet generated at depth i is $2^{k-i} + 2$ for $0 \leq i \leq k - 1$, where the constant 2 accounts for its regeneration from its two children. Regeneration requires $2^k - 1$ mix-split steps. Hence, the total number of mix-split operations will be:

$$\begin{aligned} & 2^k + \frac{\sum_{i=0}^{k-1} 2^{k-i} \times 2^i}{2} + 2^k - 1 \\ =& 2^k + \frac{\sum_{i=0}^{k-1} 2^k}{2} + 2^k - 1 \\ =& 2^{k-1}(k + 4) - 1 \end{aligned}$$

The fact that no waste droplet is generated in this process follows easily by counting the number of input droplets (Lemma 4.2.2) and the output droplets. □

Observation 4.2.3 The *CF*-values of the gradient excluding the two boundary-*CF*s appear at the output of the generator in conformity with the post-order traversal sequence of LDT.

The following theorem provides an upper bound on the storage requirement during the generation of linear gradient.

Theorem 4.2.4 Algorithm 2 requires at most $2k$ storage electrodes at any instant of time for generating a linear dilution-gradient ($\mathcal{L}$) of size $2^{k+1}+1$.

Proof. We prove the theorem using induction on k.

Basis: For $k = 1$, $|\mathcal{L}| = 2^{1+1} + 1 = 5$. One needs to generate three *CF*s from two boundary-*CF*s. It is easy to check that we require at most two intermediate storage elements in this case. Hence, the theorem is true for $k = 1$.

Inductive hypothesis: Assume the statement is true for $k - 1$.

Inductive steps: Consider the target set $\mathcal{L}$ of size $2^{k+1} + 1$. One can split $\mathcal{L}$ into three parts: $\mathcal{L}_{\text{left}}$ that contains the first $2^k + 1$ targets of $\mathcal{L}$, i.e., $\mathcal{L}_{\text{left}} = \{\mathcal{L}_1, \ldots, \mathcal{L}_{2^k+1}\}$; $\mathcal{L}_{\text{median}}$ that contains the median target of $\mathcal{L}$; and $\mathcal{L}_{\text{right}} = \{\mathcal{L}_{2^k+1}, \ldots, \mathcal{L}_{2^{k+1}+1}\}$ (see Figure 4.2). By inductive hypothesis, the left subtree requires $2(k-1)$ storage. Additionally, we need to store one droplet of *CF* ($\mathcal{L}_{2^k+1}$) corresponding to the root. So, a total of $2(k-1)+1 = 2k-1$ storage is required in order to generate all the *CF*s on the left subtree.

When we generate the target set for the right subtree, we need to store the root *CF* of the left subtree for regeneration purpose. By an analogous argument, we can claim that the right subtree requires $2k-1$ storage. Hence, the total number of storage required is $(2k-1) + 1 = 2k$. □

A layout architecture and the dynamics of linear gradient generation algorithm can be found in [LDT].

4.3 Exponential Gradient

In this section, we first discuss an *REMIA*-based [HLC12] dilution algorithm that scans the binary representation of a target-*CF* left to right. The dilution algorithm uses exponential-*CF*s as input to achieve the desired target-*CF*. Note that set of all primary exponential-*CF*s, i.e., $\{\frac{2^d}{2^d}, \frac{2^{d-1}}{2^d}, \cdots, \frac{1}{2^d}, \frac{0}{2^d}\}$, where the elements are in geometric progression, can be easily produced from the raw sample of $CF = \frac{2^d}{2^d}$ by serially mixing it with the buffer solution of $CF = \frac{0}{2^d}$. Afterwards, some underlying combinatorial properties of the *CF*s are utilized favorably for producing a monotonic sequence of *CF*s over a broad range of exponential gradients [BBH⁺13a]. Let us briefly revisit the method of diluting a sample to a desired *CF*.

Assume that all primary exponential-*CF*s, i.e., $\left\{\frac{2^d}{2^d}, \frac{2^{d-1}}{2^d}, \cdots, \frac{1}{2^d}, \frac{0}{2^d}\right\}$ are available. Algorithm 3, described below, is built based on *REMIA* [HLC12]; it outputs a wider class of exponential *CF*-values and the corresponding mix-split sequence required for generating a particular target-*CF* while scanning the bits in its binary representation, from left to right.
The following example illustrates the generation of target-$CF = \frac{847}{2^{10}}$ starting from sample ($CF = \frac{2^{10}}{2^{10}}$) and buffer ($CF = \frac{0}{2^{10}}$), i.e., for $d = 10$.

Example 4.3.1 The 10-bit binary representation of the numerator of $\frac{847}{2^{10}}$ is $(1101001111)_2$. Algorithm 3 scans the binary string from left to right and outputs the required exponential-*CF*s: $\left(\frac{2^{10}}{2^{10}}, \frac{2^{10}}{2^{10}}, \frac{512}{2^{10}}, \frac{128}{2^{10}}, \frac{128}{2^{10}}, \frac{128}{2^{10}}, \frac{64}{2^{10}}\right)$ in array variable 'exp_gradient'; they should be mixed sequentially in reverse order to reach the target. The corresponding mixing tree of Figure 4.3 shows how the target-*CF* is generated. The black dot in the figure represents a waste droplet left out of each of the intermediate mix-split steps. ■

Note that exponential-*CF*s produced by Algorithm 3 are monotonically decreasing. Furthermore, *CF*s in the array are used in the reverse order (by scanning high-index to low-index elements of 'exp_gradient') for generating the mixing tree corresponding to the target-*CF*. Hence, the sequence

Algorithm 3: Exponential gradients and mixing sequence for a target

```
Input: Target concentration (CF = x/2^d)
Output: Exponential CF-values for generating target-CF
Represent the x as d-bit binary number, i.e., x = (x_{d-1}x_{d-2}...x_1x_0)_2 ;
index = 0;
// 'index' is used for indexing an array of exponential-CFs i.e., 'exp_gradient'
level = 1;
foreach i from d - 1 down to 0 do
    if x_i is 1 then
        exp_gradient[index] = 2^i × 2^level;
        level = level + 1;
        index = index + 1;
exp_gradient[index - 1] = exp_gradient[index - 1]/2;
Execute (1:1) mix-split operations sequentially in reverse order of "exp_gradient" to generate the
target;
```

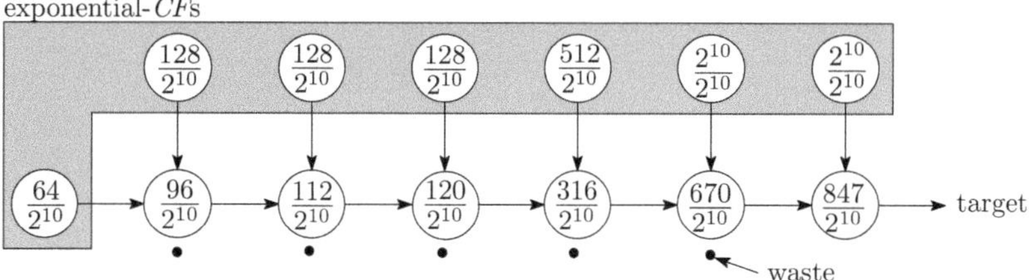

Figure 4.3 Illustrating the generation of target $CF = \frac{847}{2^{10}}$.

of exponential-*CF*s used in mix-split steps of the dilution process is monotonically increasing. This leads to the following useful observation, stated as follows.

Observation 4.3.1 When Algorithm 3 is run, the target-*CF* is reached following a strict monotonically increasing sequence of concentration factors.

The *CF*-values of waste droplets generated in the mix-split steps of the dilution process will thus satisfy the monotonicity property as stated in Observation 4.3.1. We utilize this fact to generate a wide range of exponential concentration gradients.

Initially, the primary exponential gradient, i.e., $\left\{\frac{2^d}{2^d}, \frac{2^{d-1}}{2^d}, \cdots, \frac{1}{2^d}, \frac{0}{2^d}\right\}$ can be generated fully or in part, by diluting the raw sample ($CF = \frac{2^d}{2^d}$) with the buffer ($CF = \frac{0}{2^d}$) serially. At every step of dilution, *CF*-value is halved with respect to that of the previous gradient. The following discussion shows how a more general class of exponential gradients can be generated by mixing droplets of different initial concentrations with other *CF*-values successively. In each mixing step, the same *CF* (other than buffer) can also be used. It may be noted that, if one repeatedly mixes a droplet with a *CF*, which is higher (lower) than the starting *CF*, then a monotonically increasing (decreasing) exponential gradients will be generated. Let us consider an illustrative example.

Example 4.3.2 If we start from $\frac{64}{2^{10}}$ and mix it with $\frac{128}{2^{10}}$ repeatedly, we obtain a gradient $\left\{\frac{64}{2^{10}}, \frac{96}{2^{10}}, \frac{112}{2^{10}}, \frac{120}{2^{10}}, \frac{124}{2^{10}}, \frac{126}{2^{10}}, \frac{127}{2^{10}}\right\}$. Figure 4.4 shows the plots of different exponential gradients, which are produced from an initial *CF* of $\frac{64}{2^{10}}$ by mixing it successively with $\frac{256}{2^{10}}, \frac{128}{2^{10}}, \frac{32}{2^{10}}$, and $\frac{16}{2^{10}}$.

Theorem 4.3.1 helps us to characterize all such gradients. Beforehand we need to define a *differential sequence* of a gradient set as follows.

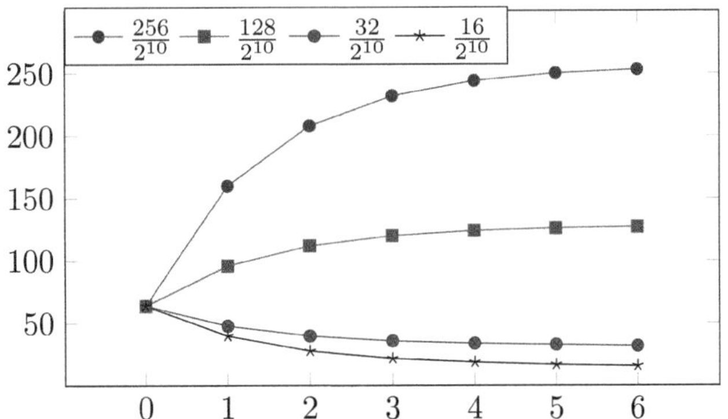

Figure 4.4 Different exponential gradients ($d = 10$).

Define the *differential sequence* of a gradient set $T = \{t_1, t_2, t_3, \cdots, t_{m-1}, t_m\}$ as a new sequence $T_\Delta = \{|t_1 - t_2|, |t_2 - t_3|, \cdots, |t_{m-1} - t_m|\}$. For example, the differential sequence of $\{\frac{64}{2^{10}}, \frac{96}{2^{10}}, \frac{112}{2^{10}}, \frac{120}{2^{10}}, \frac{124}{2^{10}}, \frac{126}{2^{10}}, \frac{127}{2^{10}}\}$ is $\{\frac{32}{2^{10}}, \frac{16}{12^{10}}, \frac{8}{2^{10}}, \frac{4}{2^{10}}, \frac{2}{2^{10}}, \frac{1}{2^{10}}\}$.

Theorem 4.3.1 If a sample droplet with *CF* $C_1 = \frac{x}{2^d}$ is repeatedly mixed with another droplet with *CF* $C_2 = \frac{y}{2^d}$, following a sequence of (1:1) mix-split steps, then the differential sequence of the generated dilution gradient will be in a geometric progression with initial term $\frac{|x-y|}{2^{d+1}}$, last term $\frac{1}{2^d}$, and with common ratio $\frac{1}{2}$, where x and y are positive integers, and $x \neq y$.

Proof. Let a unit-volume droplet of *CF* $= C_1$ be repeatedly mixed with a unit-volume droplet with *CF* $= C_2$. There are two cases: (i) $C_2 > C_1$, for which the resulting gradient will be monotonically increasing and (ii) $C_2 < C_1$, which results in a monotonically decreasing gradient. If $C_1 = C_2$, then there will be no change in gradient value. Also proving only Case (i) is sufficient, the other case can be proven similarly. For simplicity, the denominator of *CF*-values are ignored.

The resulting *CF*-sequence is: $T = \{x, \frac{x}{2} + \frac{y}{2}, \frac{x}{4} + \frac{y}{4} + \frac{y}{2}, \cdots, \frac{x}{2^k} + \sum_{j=1}^{k} \frac{y}{2^j}\}$. Note that every element in T has 2^d as the denominator, which is ignored for simplicity. Denote the ith element in T and its differential sequence T_Δ as T^i and T^i_Δ, respectively. Clearly, the initial term of the differential sequence is $\frac{1}{2^d} \times \frac{|x-y|}{2} = \frac{|x-y|}{2^{d+1}}$ (2^d is in the denominator).

Now $T^i = \frac{x}{2^i} + \sum_{j=1}^{i} \frac{y}{2^j}$ and $T^{i+1} = \frac{x}{2^{i+1}} + \sum_{j=1}^{i+1} \frac{y}{2^j}$. So,

$$T_{\Delta}^{i} = T^{i+1} - T^{i}$$

$$T_{\Delta}^{i} = \frac{x}{2^{i+1}} + \sum_{j=1}^{i+1} \frac{y}{2^j} - \frac{x}{2^i} - \sum_{j=1}^{i} \frac{y}{2^j}$$

$$= \frac{y}{2^{i+1}} - \frac{x}{2^{i+1}} = \frac{1}{2^i} T_{\Delta}^{0}$$

The last term of differential sequence can be rounded off as $\frac{1}{2^n}$, as in each step, the previous term is halved.

Hence, the differential sequence (T_{Δ}) of the generated dilution gradient (T) is in geometric progression with initial term $\frac{|x-y|}{2^{d+1}}$, last term $\frac{1}{2^d}$, and with common ratio $\frac{1}{2}$, where x and y are positive integers, and $x \neq y$. □

4.4 Complex-shaped Gradients

Gradients having complex shapes such as parabolic, sinusoidal, or Gaussian also play important roles in biochemical analyzes [DCJW01, JHK+11, LKA+09]. In this section, we present a new and generalized gradient-generation scheme on DMFB that approximates any complex-shaped gradient profile with a sequence of linear gradients; each of them can then be processed separately using the earlier technique for generating linear dilution gradient (Section 4.2). Various methods have been proposed in the literature [LCG+13] for approximating a nonlinear function with piecewise linear segments. Unfortunately, these methods do not account for the inherent error that appears while approximating a given target-*CF* depending on the user-defined error-tolerance $\epsilon, 0 \leq \epsilon < 1$. Also, the impact of the underlying mixing model supported by a DMFB platform cannot be incorporated directly in the linearization procedure. In order to handle this problem, we have used certain digital-geometric techniques for identifying digital straight line segments from a digital representation of the gradient profile. We consider a scenario from a different discipline of image analysis called digital geometry [Ros74, BB07]. Let a real curve ($\mathcal{R}$) be digitized on a uniform square-grid, where the consecutive grid lines are separated by unit distance. In the digitized version ($\mathcal{D}(\mathcal{R})$) of the curve, which consists of a sequence of grid-points, a grid-point p is assigned to $\mathcal{D}(\mathcal{R})$, if the distance $\|p, \alpha\| \leq 0.5$,

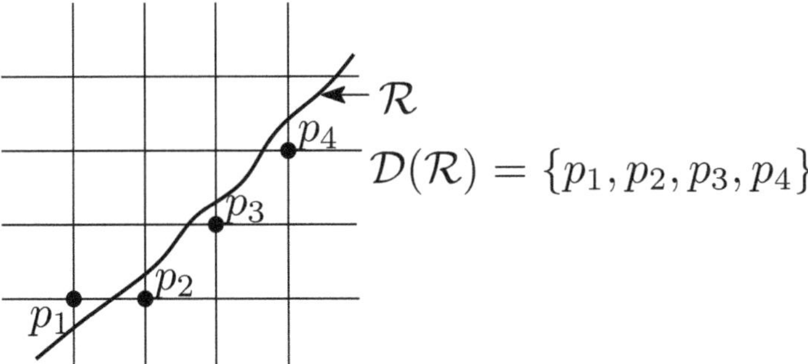

Figure 4.5 Digitization [Ros74] of real curve $\mathcal{R}$ as a sequence of grid-points on a unit-square grid.

where α is the intersection point of $\mathcal{R}$ and the vertical line passing through p [Ros74]. The grid-points correspond to pixels in the planer-dual space of grid graphs. Figure 4.5 shows a real curve $\mathcal{R}$ and its digitization, i.e., $\mathcal{D}(\mathcal{R})$, as a sequence of grid-points. Observe that in the context of sample preparation, $\frac{1}{2^{d+1}} = \frac{0.5}{2^d}$. As is apparent from the grid-point (p) assignment rule (corresponding to the intersection (α) of a real curve and a vertical grid line) and the aforementioned observation, the same approximation policy is adopted for both representing target-*CF* in sample problem and choosing grid-points for digital representation of a curve. Before going to the detailed description of the complex-shaped gradient generation method, few useful definitions related to this context are given as follows.

Definition 4.4.1 *(Chain code)* Let $p = (x, y)$ be a grid-point. A grid-point (x', y') is a neighbor of p, if $\max\{|x - x'|, |y - y'|\} = 1$. The chain code [Fre61] of p with respect its neighboring grid-point in a digital curve can have a value in $\{0, 1, 2, \ldots, 7\}$, assuming 8-connectivity of the underlying digital curve.

Definition 4.4.2 A digital curve (DC) is an ordered sequence of grid-points (representable by chain code) such that each point (excepting the first one) in the digital curve is a neighbor of its predecessor in the sequence.

Figure 4.6(a) shows the chain code for 8-neighbors around a grid-point. Additionally, Figure 4.6(b) shows the grid-points of a digital curve and its chain code starting from the grid-point (1,2).

Definition 4.4.3 A digital curve C has the chord property [Ros74] if, for every grid-points p, q in C, the chord $\overline{pq}$ (the line segment joining p and q in real

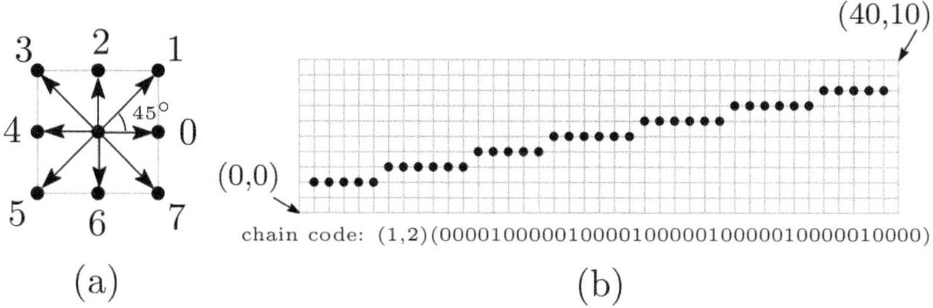

Figure 4.6 (a) Chain code for 8-neighbors around a grid-point, (b) digital curve as a sequence of grid-points on a unit-square grid.

plane) "lies near" C, i.e., for any point (x, y) of $\overline{pq}$, there exists some point (x', y') of C such that $\max\{|x - x'|, |y - y'|\} < 1$.

It has been shown by Rosenfeld that a digital curve (C) is the digitization of a straight line segment if and only if C satisfies the chord property [Ros74]. The properties of a digital straight line segment (DSS) are listed as follows:

(P1) The runs have at most two directions, differing by $45°$, and for one of these directions, the run length must be 1.
(P2) The runs can have only two lengths, which are consecutive integers.
(P3) One of the run lengths can occur only once at a time.
(P4) For the run length that occurs in runs, these runs can themselves have only two lengths, which are consecutive integers; and so on.

Example 4.4.1 In Figure 4.6(b), the chain code of the digital curve is $0^4 10^5 10^4 10^5 10^5 10^5 10^4$. Note that the directions of 0 and 1 in the chain code differ by $45°$ (Figure 4.6(a)), and along direction 1, the run length is 1. Hence, P1 is satisfied. The run length for 0 in the chain code is 4545554, and it has two consecutive lengths (4 and 5) only; hence, P2 is satisfied. In the chain code ($45^1 45^3 4$), run length of 4 is 1, and run lengths of 5 are 1 and 3. As '4' has only one run length, condition P3 is true. However, P4 is not satisfied because run lengths of 5 (i.e., 1 and 3) are not consecutive integers. ■

The digital curve shown in Figure 4.6(b) can be represented using a sequence of two DSSs having chain codes $0^4 10^5 10^4 10^5 10^5 1$ and $0^5 10^4$, respectively. Our objective is to represent the digital curve of a gradient profile using fewest DSS components. The following subsection illustrates the digital curve representation of a gradient profile depending on the user-defined error-tolerance limit $\epsilon, 0 \leq \epsilon < 1$.

4.4.1 Digital Curve Representation of a Gradient Profile

In the case of dilution, any target concentration $c_t \in [0,1]$ is approximated as $\frac{x}{2^d}$, where $x, d \in \mathbb{N}$ and $0 \leq x \leq 2^d$, depending on the user-defined error-tolerance limit ϵ, and the underlying (1:1) mixing models supported by DMFB. The *CF* of sample and buffer is represented as $\frac{2^d}{2^d}$ and $\frac{0}{2^d}$, respectively. Note that the maximum possible error in *CF* is bounded by $\frac{1}{2^{d+1}}$, i.e., $|c_t - \frac{x}{2^d}| \leq \frac{1}{2^{d+1}} < \epsilon$, where $c_t \in [0,1]$ is the desired target concentration and $\frac{x}{2^d}$ is the approximated target-*CF*.

In the process of digitizing a gradient profile, we need to sample the gradient profile in such a way that the absolute difference between any two consecutive sample points lying on the gradient profile is at most $\frac{1}{2^d}$, which in turn, ensures the connectivity of the chain code in the digitized gradient profile.

In order to demonstrate the principle with an example, assume that we need to generate *CF*s that follow a parabolic gradient profile satisfying user-specified error-tolerance limit $\epsilon = 0.002$ on a DMFB implementing (1:1) mixing model. Let us consider a parabolic gradient profile $y = f(x) = x^2$ for $x \in [0,1]$. Note that if we set $d = 8$, the error-tolerance limit for each approximated *CF* in $\left\{\frac{1}{2^8}, \frac{2}{2^8}, \frac{3}{2^8}, \cdots, \frac{2^8-1}{2^8}\right\}$ is satisfied because $\frac{1}{2^9} = 0.0019 < 0.002$. The entire range of y-axis ($y \in [0,1]$ for $x \in [0,1]$ on $y = x^2$) is partitioned into $2^8 + 1$ *CF*-values, where the absolute difference between any two consecutive points are $\frac{1}{2^8}$, i.e., each sample point, or the approximated *CF* satisfies the error-tolerance limit. Now, we need to determine the sampling distance (Δx) on the x-axis so that connectivity of the digitized gradient profile is preserved. Note that $\Delta y = f'(x)\Delta x$ and the maximum value of Δy for $x \in [0,1]$ is $2\Delta x$, i.e., Δy is maximized at $x = 1$. For $d = 8$, $\Delta y = \frac{1}{2^8}$. Hence, $\Delta x = \frac{1}{2^9}$. The digital curve representation of the parabolic ($y = x^2$) gradient profile is shown in Figure 4.7 for $\epsilon = 0.002$, and a small portion of the Figure 4.7 is zoomed into Figure 4.8. Let us denote the digital curve representation of a gradient profile as "digital gradient profile".

4.4.2 Identification of DSS on a Gradient Profile

The digitized gradient profile can be represented with a chain code. Our objective is to identify a set of linear gradients (i.e., equivalent to DSS) where the concentrations on each linear gradient belongs to the original parabolic profile. We apply the algorithm proposed in [Ros74] for finding a set of DSS from the chain code representation of a digital gradient profile. Figure 4.9 shows the fewest number of DSSs that represent the parabolic digital gradient

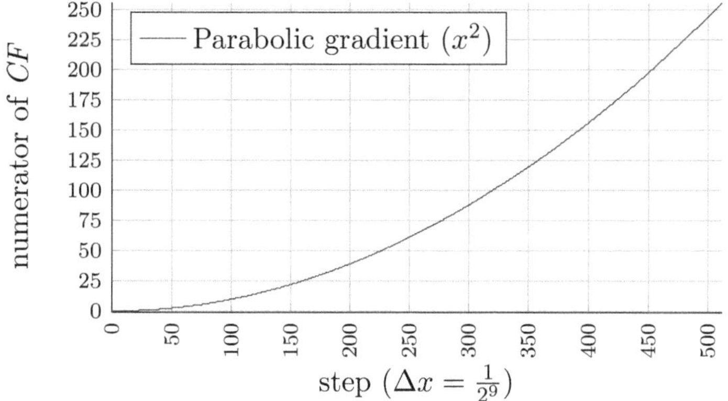

Figure 4.7 DC representation of $y = x^2$ for $\epsilon = 0.002$.

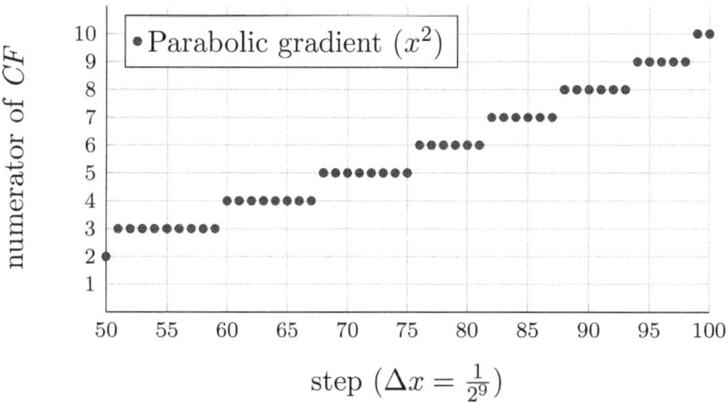

Figure 4.8 Zoomed digital curve $y = x^2$ for $\epsilon = 0.002$.

profile. As the DSS extraction algorithm proposed in [Ros74] is based on the chord property, the user-specified error-tolerance limit is satisfied for each individual *CF*s. However, one can easily observe that the number of DSS-components extracted using [Ros74] are quite large. This is because of the stringent requirement of satisfying error-tolerance in each individual *CF* lying on the gradient profile. However, such stringent requirement of error-tolerance satisfaction for each individual *CF* may not always be necessary for most of the biological protocols [DCJW01]. Hence, we can represent a gradient profile using fewer number of straight line segments compared to the number of DSSs [BB07].

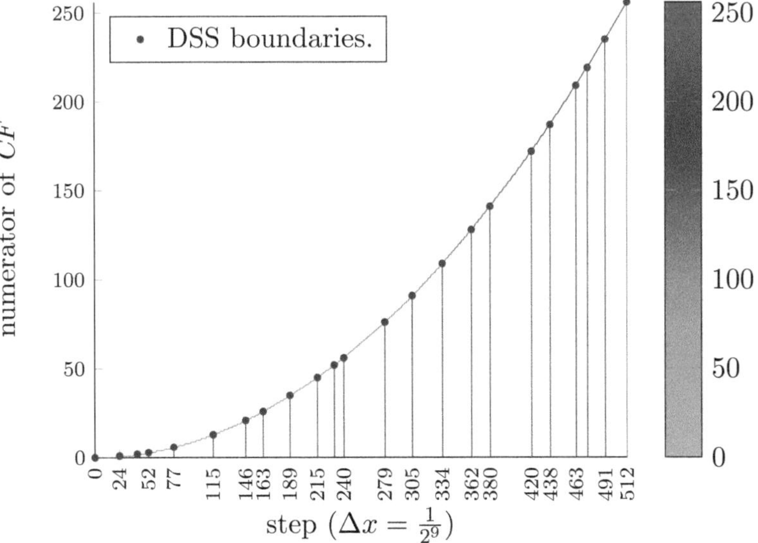

Figure 4.9 DSSs identified by [Ros74].

Bhowmick *et al.* [BB07] proposed an algorithm for finding approximate digital straight line segments (ADSS) by preserving some of the most fundamental properties of a DSS, while relaxing or dropping a few others. The algorithm proposed there retains (R1) and modifies (R2) while dropping (R3) and (R4) from the formal characterization of DSS given in [Ros74]. Modification of (R2) allows the run-length to vary by more than unity depending on the minimum run length. The algorithm for extracting ADSS as proposed in [BB07] has the following properties:

(P1) The runs have at most two directions, differing by $45°$, and for one of these directions, the run length must be 1.

(P2′) The runs can have a range of possible lengths ($[p, q]$, excepting l and r) such that

1. $q - p \leq d = \lfloor \frac{(p+1)}{2} \rfloor$
2. $(l - p), (r - p) \leq e = \lfloor \frac{(p+1)}{2} \rfloor$

where l (r) is the length of the leftmost (rightmost) run of nonsingular element in chain code, d and e represent the allowance of approximation.

Example 4.4.2 In Figure 4.6(b), the chain code of the digital curve is $0^4 10^5 10^4 10^5 10^5 10^5 10^4$, where 0 (1) is the nonsingular (singular) element

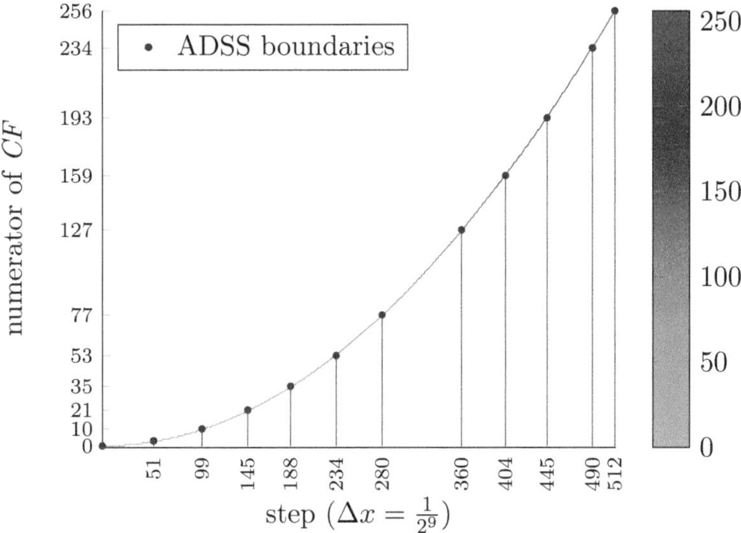

Figure 4.10 Approximate digital straight line segments (ADSSs).

in the chain code. The length of the leftmost run of nonsingular element is 4, i.e., $l = 4$. Similarly, $r = 4$. Note that, $p = 4$ and $q = 5$. Hence, the condition (P2′) is satisfied and the entire chain code is recognized as a single ADSS as opposed to two DSSs. ■

Figure 4.10 shows the ADSS segmentation of the parabolic digital gradient profile with fewest number of ADSS components, where an ADSS appears between two consecutive vertical lines. Note that it requires only 11 ADSSs instead of 22 DSSs for representing the digital parabolic gradient profile. In the complex-shaped gradient generation technique, we have used the ADSS algorithm [BB07] for segmenting a digital gradient profile. Also, for a gradient profile, we need to generate a set of *CF*s lying on the profile that are separated uniformly on the abscissa. We have used linear dilution gradient generation algorithm (presented in Section 4.2) for generating the *CF*s lying on each individual ADSS. The following example illustrates the gradient generation scheme.

Example 4.4.3 Assume that we need to generate 34 *CF*s lying on the parabolic gradient $y = x^2, x \in [0, 1]$. We consider 512 uniform sample points for representing parabolic gradient profile satisfying the error-tolerance limit $\epsilon = 0.002$ (Section 4.4.1). We chose 34 uniformly separated points that are

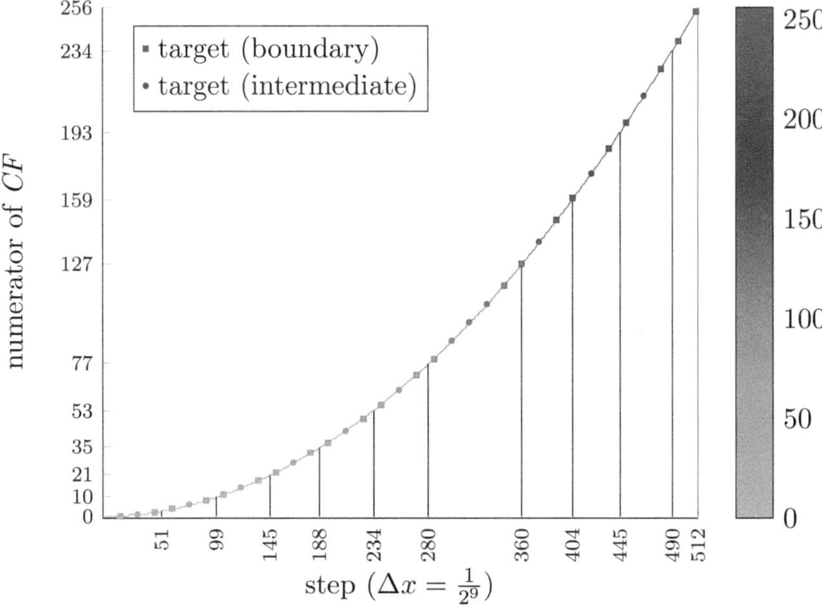

Figure 4.11 Generating 34 uniformly separated *CF*s lying on a parabolic gradient.

15 units apart ($\lfloor \frac{512}{34} \rfloor = 15$). Moreover, these target-*CF*s lie on the ADSS components that are used for representing the parabolic gradient profile. Note that all *CF*s lying on each individual ADSS form a linear gradient and hence, they can be categorized into types: boundary and intermediate *CF*s, where boundary-*CF*s are used for generating intermediate *CF*s. Figure 4.11 shows these two types of *CF*s and it can be observed that out of 34 target-*CF*s, 22 boundary-*CF*s need to be generated using sample and buffer for producing the remaining 12 intermediate target-*CF*s. ■

Other gradient profiles: We have also considered sinusiodal and Gaussian gradient profiles [LKA$^+$09, WJWL10] as shown in Figure 4.12 and Figure 4.13, respectively, and used the same technique to decompose them into a set of linear gradients.

In order to produce intermediate *CF*s for a linear gradient lying on an ADSS, we need to generate boundary *CF*s (each having a fixed demand of a certain number of droplets) using sample and buffer. We have used a dilution engine [RBGC14a] for producing multiple droplets of a target-*CF* as needed.

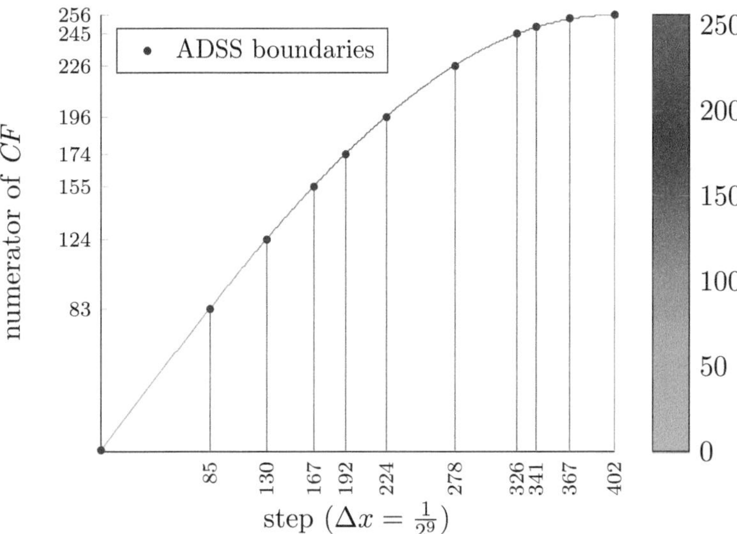

Figure 4.12 Approximate digital straight line segments (ADSSs) on a sinusiodal gradient.

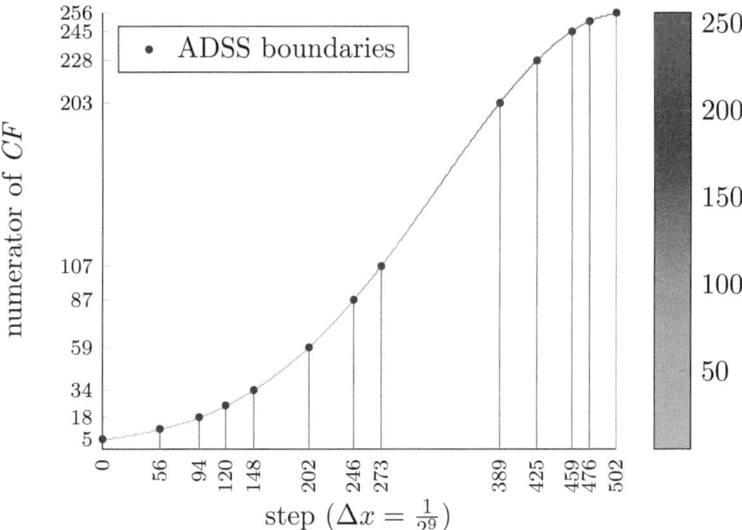

Figure 4.13 Approximate digital straight line segments (ADSSs) on a Gaussian gradient.

4.5 Experimental Results

We compare the new gradient generation technique with other general-purpose sample-preparation algorithms such as *RSM* [HHC14], *REMIA* [HLC12], and *WARA*[1] [HLL13] for generating linear, exponential, parabolic, and few other commonly used gradients. The gradient generation algorithm for complex shapes is based on the linear-gradient method as we have approximated them with piecewise-linear gradients. We report the average value of each performance parameter with a bar chart. However, in the case of other gradients, we have considered a few gradient profiles, and for each of them, we present the results in tabular form.

4.5.1 Linear Gradient

In our experiments, we have considered six different linear dilution gradient sets $\mathcal{L}$ ($|\mathcal{L}| = 2^k + 1$) for $k = 2, \ldots, 7$. For each k, we have chosen 100 random sets in the range of $\frac{1}{1024}$ and $\frac{1023}{1024}$, assuming $d = 10$. Hence, the error in concentration factors will be at most $\frac{1}{2048}$. Assuming 100% sample and 0% buffer as the two boundary-*CF*s, we have counted the total number of (1:1) mix-split steps and waste droplets using a dilution engine [RBGC14a] as well as the gradient generator. Comparative results with respect to earlier methods are shown as histograms in Figure 4.14 and Figure 4.15, where the horizontal axis indicates the size of the target set ($|\mathcal{L}| = 2^k + 1$) for $k = 2, \ldots, 7$, and the vertical axis represents the average number of mix-split steps and waste droplets required in these methods. Note that the waste droplets that are generated correspond to those produced by the dilution engine while preparing boundary-*CF*s. We observe that our method produces fewer number of waste droplets compared to *RSM* [HHC14], *REMIA* [HLC12], and *WARA* [HLL13] for $2 \leq k \leq 7$. Furthermore, the method performs better in terms of the number of mix-split steps up to a target set of size 65, i.e., up to $k = 6$.

We have evaluated the performance of the gradient generation method from the perspective of reactant usage as well by comparing it with *REMIA* [HLC12] and *WARA* [HLL13]. Figure 4.16 and Figure 4.17 show the histogram of average sample and buffer consumption obtained by *REMIA*, *WARA*, and our method. It is observed that the new gradient generation method saves the reactants significantly for a target set of size 17 or more.

[1]We have considered droplet-sharing feature only while implementing.

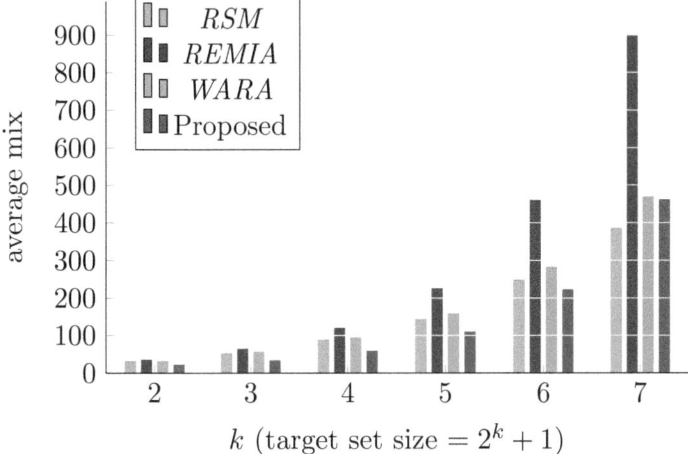

Figure 4.14 Average number of mix-split steps.

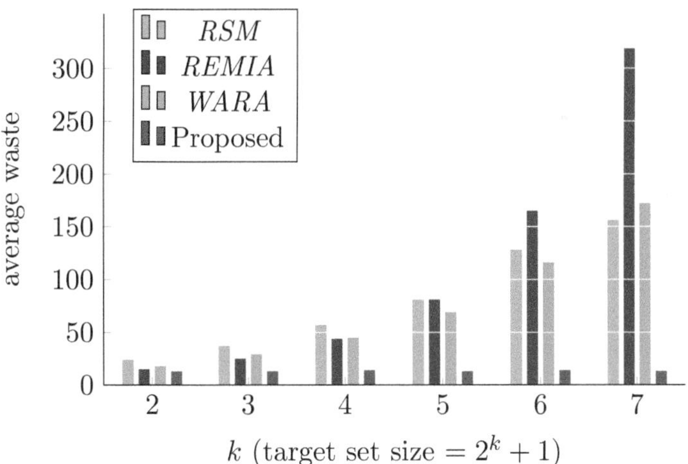

Figure 4.15 Average number of waste droplets.

4.5.2 Exponential Gradients

We have compared the performance of the gradient generation technique on several exponential gradients generated with generalized sample preparation algorithms *RSM* [HHC14], *REMIA* [HLC12], and *WARA* [HLL13]. In our study, we have chosen $d = 10$; thus, each *CF* is represented as $\frac{x}{2^{10}}$, where $0 \leq x \leq 1024$. Comparative results are shown in Table 4.1, where only the numerator parts of concentration factors are shown for convenience.

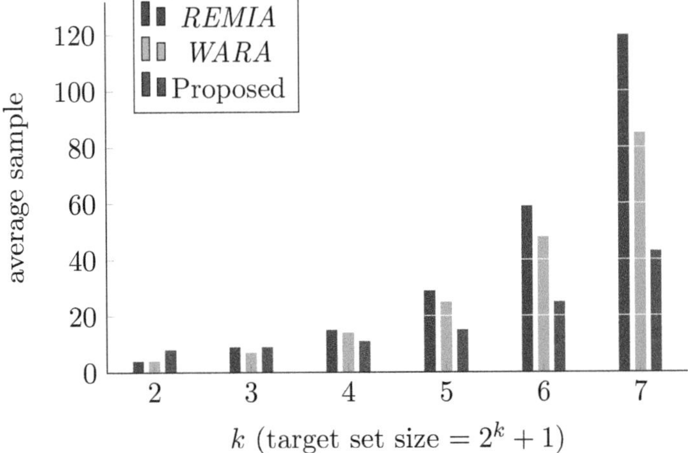

Figure 4.16 Average number of sample droplets.

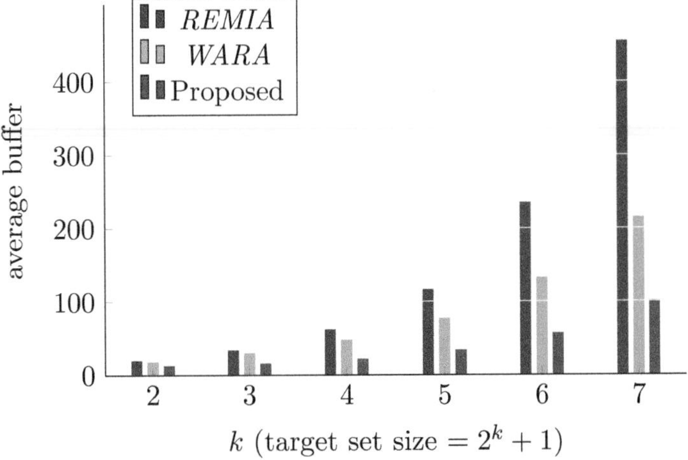

Figure 4.17 Average number of buffer droplets.

It may be observed that for most of gradient sets considered in our experiment, the new gradient generation method performs better with respect to the number of mix-split (n_m) operations, waste (n_w) droplets, and both the sample (n_s) and buffer (n_b) consumption compared to *RSM* [HHC14], *REMIA* [HLC12], and *WARA* [HLL13].

Table 4.1 Experimental results on exponential gradients

Target set	Proposed				RSM [HHC14]				REMIA [HLC12]				WARA[a][HLL13]			
$d = 10$	n_m	n_w	n_s	n_b	n_m	n_w	n_s	n_b	n_m	n_w	n_s	n_b	n_m	n_w	n_s	n_b
64, 160, 208, 232, 244, 250, 253	14	3	2	8	22	16	7	16	45	23	6	24	14	3	2	8
128, 192, 224, 240, 248, 252, 254, 255	14	1	2	7	24	17	8	17	56	28	8	28	14	1	2	7
512, 320, 224, 176, 152, 140, 134, 131	15	2	2	8	29	23	10	21	34	17	4	21	34	17	4	21
256, 640, 832, 928, 976, 1000, 1012, 1018, 1021	10	2	9	2	10	2	9	2	44	38	39	8	10	2	9	2
1024, 768, 640, 576, 544, 528, 520, 516, 514, 513	14	2	7	5	18	10	11	9	18	10	11	9	18	10	11	9

[a]only droplet-sharing part has been implemented.

n_m: the number of mix-split steps, n_w: the number of waste droplets, $n_s(n_b)$: the number of sample (buffer) droplets.

4.5.3 Parabolic, Sinusoidal, and Gaussian Gradients

In our experiments, we consider other gradient profiles of different shapes (parabolic, sinusoidal, and Gaussian) and generate *CF*s that are separated uniformly on the abscissa of the gradient profile. Note that each *CF* lying on a gradient profile is represented as $\frac{x}{2^8}, 0 \leq x \leq 2^8$, i.e., the approximation error in *CF* is less than or equal to $\frac{1}{2^9}$. In the digitization process (Section 4.4.1), we have taken $512, 402$, and 502 uniformly separated sample points for parabolic ($x^2, x \in [0,1]$), sinusoidal ($\sin x, x \in [0, \frac{\pi}{2}]$), and Gaussian ($e^{-(x-2)^2}, x \in [0,2]$) gradient profiles, respectively. Table 4.2 shows comparative results when the target *CF*s lie on the entire range of the desired gradient profile; for $d = 8$, the numerator of a target *CF* lies in the range $[0, 2^8]$. The new algorithm outperforms *RSM* with respect to the number of waste, sample, buffer droplets, and beats *REMIA* with respect to all parameters. In the case of *WARA*, it performs better with respect to the number of mix-split steps, waste, and buffer droplets and compares favorably in terms of the number of sample droplets. We have also computed mean square error (MSE) between the approximated and those generated with the new gradient generation method (Table 4.2). The value of MSE is found to be small for each of these test cases. As the existing literature suggests [KLK$^+$08, LKA$^+$09], up to 5% average concentration variation of *CFs* is acceptable on a gradient profile.

In most of the real-life test cases, target *CF*s lying on a specific portion of gradient profile are required. Table 4.3 shows such comparative results. It may be observed that the performance of the new gradient generation method becomes significantly better compared to earlier methods [HHC14, HLC12, HLL13]. This is due to the decrease in the number of linear segments required for representing the desired portion of the gradient profile. This is also evident from the experimental results reported in Section 4.5.1. The *CF*-values corresponding to different gradient profiles (parabolic, sinusoidal, and Gaussian) are shown in Figure 4.11-4.13. The number of linear segments in each test case can be determined from the respective gradient profile. As expected, the performance of the gradient generation technique improves with the reduction in the number of ADSS-components that are necessary to segment the gradient profile.

Table 4.2 Experimental results on parabolic, sinusoidal, and Gaussian gradients, where target CFs span the entire range of CFs.

Gradient	Size	Proposed					*RSM* [HHC14]				*REMIA* [HLC12]				*WARA*[a][HLL13]			
		n_m	n_w	n_s	n_b	MSE	n_m	n_w	n_s	n_b	n_m	n_w	n_s	n_b	n_m	n_w	n_s	n_b
Parabolic	34	93	51	34	51	0.12	95	65	42	57	152	82	41	75	119	59	37	56
	50	131	53	41	62	0.42	120	74	47	76	233	125	60	113	158	73	50	72
Sinusoidal	25	77	44	43	26	0.04	67	45	42	28	140	92	65	52	84	46	42	29
	40	105	48	54	34	0.16	99	63	61	42	212	136	97	79	126	65	60	45
Gaussian	33	99	57	41	49	0.05	99	68	44	57	163	93	53	73	118	57	39	51
	50	147	72	56	66	0.11	132	85	58	77	254	145	83	112	166	76	58	68

[a]only droplet-sharing part has been implemented.

n_m: the number of mix-split steps, n_w: the number of waste droplets, $n_s(n_b)$: the number of sample (buffer) droplets.

Table 4.3 Experimental results on Parabolic, sinusoidal, and Gaussian gradients, where target CFs span a specific portion of the CF-range.

Gradient	CF range	Size	Proposed				*RSM* [HHC14]				*REMIA* [HLC12]				*WARA*[a][HLL13]			
			n_m	n_w	n_s	n_b	n_m	n_w	n_s	n_b	n_m	n_w	n_s	n_b	n_m	n_w	n_s	n_b
Parabolic	[10, 50]	15	49	23	8	30	49	36	14	37	67	30	5	40	52	20	4	31
	[50, 160]	25	73	27	24	28	88	66	41	50	155	83	32	76	99	45	25	45
	[155, 255]	25	70	21	33	13	83	60	58	27	145	98	73	50	99	61	58	28
Sinusoidal	[0, 150]	30	87	29	20	39	91	64	38	56	152	75	25	80	113	49	21	58
	[175, 255]	28	79	24	40	12	85	59	61	26	167	116	88	56	97	59	62	25
Gaussian	[25, 90]	25	72	22	14	33	82	59	28	56	138	66	15	76	92	35	11	49
	[110, 240]	30	84	18	30	18	96	68	55	43	170	103	64	69	113	61	49	42

[a]only droplet-sharing part has been implemented.

n_m: the number of mix-split steps, n_w: the number of waste droplets, $n_s(n_b)$: the number of sample (buffer) droplets.

4.6 Conclusions

In this chapter, we have addressed the problem of multi-target sample preparation with digital microfluidics, where the target concentration factors satisfy certain gradient patterns. We have provided efficient algorithmic solutions for the generation of linear and exponential gradients. Other gradient patterns such as parabolic, sinusoidal, or Gaussian have also been analyzed. We have discussed a zero-waste protocol based on the regeneration principle for producing linear dilution-gradients on a digital microfluidic platform. For sample preparation requiring arbitrary-shaped gradient-profile, we have used digital-geometric techniques for representing it as a sequence of linear gradients. The power of the linear-gradient generation technique can then be effectively harnessed for producing the desired gradient-profile. The new gradient generation techniques offer customized solutions to various gradient-generation tasks and can be used to save costly stock solutions and sample-preparation time compared to existing approaches. Management of various fluidic errors that might affect the accuracy of *CF*s of the sample may further be studied as an open problem.

5

Concentration-Resilient Mixture Preparation

Sample preparation plays a crucial role in almost all biochemical applications since a predominant portion of biochemical analysis time is associated with sample collection, transportation, and preparation. Many sample-preparation algorithms are proposed in the literature that are suitable for execution on programmable digital microfluidic (DMF) platforms [TUTA08, RBC10, LCH15, RCK$^+$15b, KRC$^+$13, LCLH13]. In most of the existing DMF-based sample-preparation algorithms, a fixed target ratio is provided as input and the corresponding mixing tree is generated as output. However, in many biochemical applications, target mixtures with exact component-proportions may not be needed. From a biochemical perspective, it may be sufficient to prepare a mixture in which the input reagents may lie within a range of concentration factors. In many biochemical applications, some of the input reagents in the desired target ratio are allowed to have a concentration factor (*CF*) within a certain range. For example, analyte concentrations in a ratio may often be prepared keeping its constituents within 10% of the indicated value in conformance with the guidelines provided by United States Pharmacopeia and the National Formulary [USP]. In polymerase chain reaction (PCR) with Taq DNA Ploymerase [PCRb], both forward and reverse primer concentration may vary between 0.05 and 1 μM. Moreover, for trouble shooting in PCR, magnesium salt Mg^{2+} (at a final concentration of 0.5–5.0 mM), potassium salt K+ (35–100 mM), dimethyl sulfoxide (1–10%), formamide (1.25–10%), bovine serum albumin (10–100 μg/ml), and betaine (0.5–2.5 M) are required as additional reagents [Jov].

Although any concentration between the allowable range satisfies the protocol specification from biochemical standpoint, the cost of generating a target mixture using a digital microfluidic biochip strongly depends on the actual concentration of each of its constituents. Automated production of a mixture with a certain ratio on a chip requires a well-defined sequence of droplet mix-split steps, which determine the time and cost of reactants involved in mixture preparation. Hence, choosing a suitable concentration ratio from a given range of target concentrations that meets the given optimality criteria (minimum number of mix-split steps, i.e., mixture-preparation time, or reagent-usage, i.e., cost) is important for reducing the overall sample-preparation overhead. This task was posed as an open problem in the literature [TUTA08]. A recent work based on certain heuristics addresses this problem [LCg16].

In this chapter, we present an ILP-based concentration-resilient (or variability-aware) ratio-selection method that minimizes reactant cost when *MinMix* [TUTA08] algorithm is used for mixture preparation. The ratio selection method acts as a pre-processing step that selects a mixing ratio following which any other existing sample-preparation algorithm such as *RMA* [RCK+15b], *MTCS* [KRC+13], or *CoDOS* [LCLH13] can be invoked for efficient generation of the mixture using the chosen ratio. Figure 5.1 shows the flow of concentration-resilient sample preparation on DMFB. We observe that the selected target ratio significantly reduces the number of mix-split steps and reagent-usage compared to the case for random choice of target concentration ratio within the valid ratio space.

5.1 Related Work

We discuss some relevant sample preparation algorithms for making this chapter self-contained. Most of the state-of-the-art DMFB sample-preparation algorithms start with an approximated ratio, i.e., the input ratio

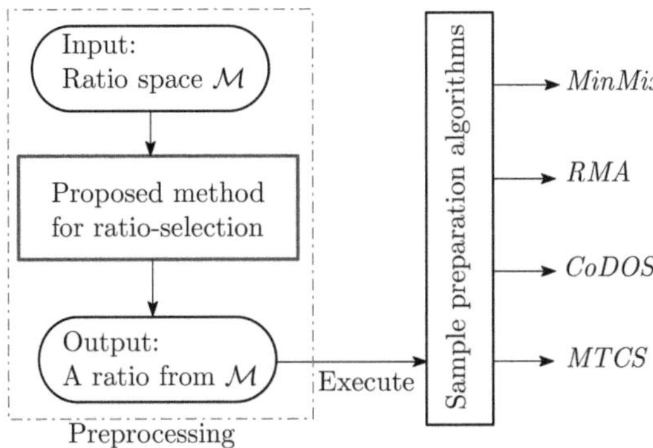

Figure 5.1 Overview of concentration-resilient sample preparation.

is of the form $\{R_1 : R_2 : \cdots : R_k\} = \{x_1 : x_2 : \cdots : x_k\}$, where $\sum_{i=1}^{k} x_i = 2^d, d \in \mathbb{N}$. Mixture preparation algorithm *MinMix* [TUTA08] performs mixing two or more input fluids with an approximated ratio $\{R_1 : R_2 : \cdots : R_k = x_1 : x_2 : \cdots : x_k\}$, where $\sum_{i=1}^{k} x_i = 2^d$, assuming (1:1) mixing model. *MinMix* algorithm represents each x_i as a d-bit binary number. After that, these k d-bit binary representations are bit-wise scanned from right to left to construct a mixing tree in bottom-up fashion. Figure 5.2(a) shows a mixing tree for the approximated target ratio $\{R_1 : R_2 : R_3 = 6 : 7 : 3\}$. Note that for each non-zero bit in the d-bit binary representation of x_i, a droplet of input reagent R_i is required as a leaf node in the *MinMix* tree.

Roy *et al.* proposed a mixing algorithm, namely Ratioed Mixing Algorithm (*RMA*) [RCK+15b], which determines a mixing tree by decomposing the algebraic expression corresponding to a target ratio in top-down fashion. For a desired target ratio, *RMA* determines a mixing tree with longer subsequences of mixing steps with a small number of distinct fluids (called "dilution subtrees"[1]). Figure 5.2(b) shows a mixing tree for $\{R_1 : R_2 : R_3 = 6 : 7 : 3\}$, in which two dilution subtrees are highlighted. Note that the mixing graph produced by *RMA* can be implemented on a chip with a fewer number of crossovers among droplet-routing paths as well as with a reduced reservoir-to-mixer transportation distance [RCK+15b]. However, the number of mixing and reagent consumption may be greater or equal to those needed in *MinMix* algorithm.

Several other mixture-preparation algorithms, e.g., common dilution operation sharing (*CoDOS*) [LCLH13] and mixing tree with common subtree

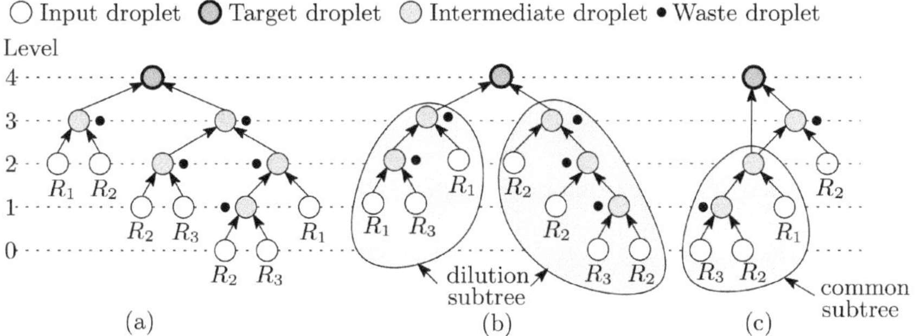

Figure 5.2 Mixing graph for the approximated target ratio $\{R_1 : R_2 : R_3 = 6 : 7 : 3\}$ using (a) *MinMix* [TUTA08], (b) *RMA* [RCK+15b], and (c) *MTCS* [KRC+13].

[1]A dilution subtree is a mixing tree in which the number of distinct leaf nodes is exactly 2.

(*MTCS*) [KRC+13], start with the mixing tree generated by MinMix, and systematically morph it to create a mixing graph based on sharing of common subtrees; such transformation reduces the number of mix-split operations, reactant cost, and waste droplets. Figure 5.2(c) shows a mixing tree generated by *MTCS*, where the common subtree at level 2 is shared with the target node at level 4. Several other sample-preparation algorithms are known in the literature [RBC10, LCH15, DYHH13] that take care of single- and multiple-target dilution problem, i.e., considering only two input reagents. In this chapter, we address the more general problem of variability-aware ratio-selection for mixture preparation with multiple input reagents.

5.2 Motivation and Problem Definition

The number of leaf nodes in the *MinMix* [TUTA08] tree, i.e., input reagents, is equal to the total number of ones present in the d-bit binary representation of the *CF* of each input reagent. The number of (1:1) mix-split steps, i.e., internal nodes, is one less than the total number of leaf nodes. Hence, sample preparation time (number of mix-split steps) and cost (reagent units) are determined from the number of non-zero bits in the target ratio in the case of *MinMix* algorithm. However, other sample preparation algorithms such as *CoDOS* [LCLH13] and *MTCS* [KRC+13] transform the mixing tree generated by *MinMix* [TUTA08]; the performance (mixing time and cost) of the *RMA* [RCK+15b] algorithm is closely related to the performance of *MinMix* algorithm. Hence, any improvement in the performance of *MinMix* is likely to enhance the performance of other related sample preparation algorithms (*MTCS, CoDOS*) that essentially morph the *MinMix* tree systematically to produce mixing graphs.

As discussed earlier, in many biochemical applications, the concentration of an input reagent in the desired target ratio may take a value within an allowable range. Any mixing ratio in which *CF* of each input reagent lies within an allowable range can be safely used in the biochemical application. Let us consider a motivating example of a mixture of four input reagents where the concentrations of R_1, R_2, R_3, and R_4 lie within the concentration range $[0.3, 0.4]$, $[0.25, 0.35]$, $[0.20, 0.25]$, and $[0.05, 0.10]$, respectively. Note that the set of valid ratio is represented as $\{R_1 : R_2 : R_3 : R_4 = c_1 : c_2 : c_3 : c_4\}$, where $0.3 \leq c_1 \leq 0.4$, $0.25 \leq c_2 \leq 0.35$, $0.20 \leq c_3 \leq 0.25$, $0.05 \leq c_4 \leq 0.10$; also solution will exist only if there exist a $c_i, \forall i$ such that $\sum_{i=1}^{4} c_i = 1$. If we choose the maximum depth of the mixing tree $d = 8$, *CF*s of R_1, R_2, R_3, and R_4 fall within the range $[\frac{77}{2^8}, \frac{102}{2^8}]$, $[\frac{64}{2^8}, \frac{89}{2^8}]$, $[\frac{52}{2^8}, \frac{64}{2^8}]$, and $[\frac{13}{2^8}, \frac{25}{2^8}]$, respectively. For simplicity, let us ignore the denominator of the fractional upper and lower bounds of the concentration range. Hence, the set of valid ratio becomes $\{R_1 : R_2 : R_3 : R_4 = c'_1 : c'_2 : c'_3 : c'_4\}$, where $77 \leq c'_1 \leq 102$, $64 \leq c'_2 \leq 89$, $52 \leq c'_3 \leq 64$, $13 \leq c'_4 \leq 25$ and $\sum_{i=1}^{4} c'_i = 2^8$. Note that the mixing tree shown in Figure 5.3 for the ratio $\{95 : 87 : 59 : 15\}$ represents a valid mixing ratio, for which the total number of mix-split steps and reagents droplets are 19 and 20, respectively. The symbol $n_m(n_w)$ denotes the number of (1:1) mix-split steps (waste droplets) needed (produced) by the mixing graph. However, $\{96 : 80 : 64 : 16\}$ also represents a valid mixing ratio that requires only five mix-split steps and six reagent droplets. The corresponding mixing tree is shown in Figure 5.4(a). Figure 5.4(b) shows the target DMFB platform, on which a sequence of mixing operations can be performed for producing the desired target ratio. Note that in automatic sample preparation of the desired target ratio $\{R_1 : R_2 : R_3 : R_4 = 96 : 80 : 64 : 16\}$ on the DMFB platform, two 1X droplets of R_2 and R_4 are mixed using (1:1) mixing module available on the target DMFB. After mixing, the 2X droplet is split into two equal volume (1X) droplets. One of them is transported into waste reservoir

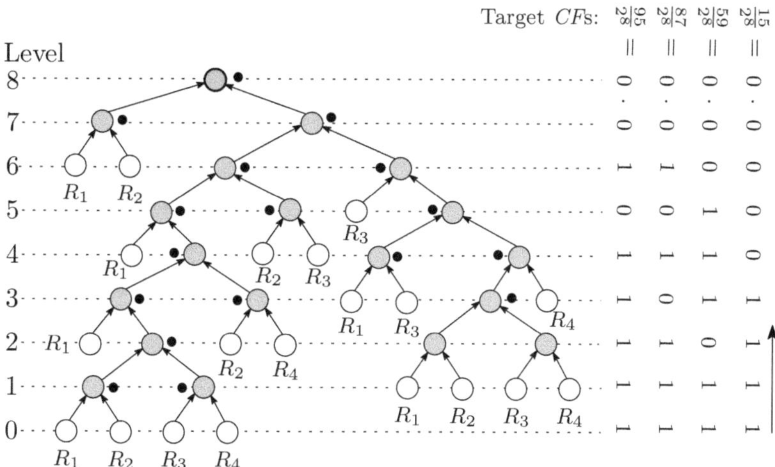

Figure 5.3 Mixing tree for the target ratio $\{R_1 : R_2 : R_3 : R_4 = 95 : 87 : 59 : 15\}$ produced by *MinMix*, where n_m, n_w number of (1:1) mix-split steps and waste droplets; n_{r_i} number of droplets of reagent R_i, for $i = 1, 2, 3, 4$.

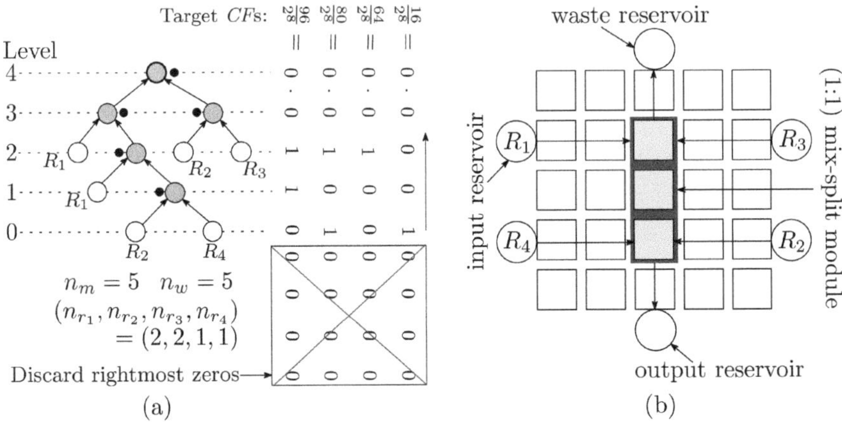

Figure 5.4 (a) Mixing tree for the target ratio $\{R_1 : R_2 : R_3 : R_4 = 96 : 80 : 64 : 16\}$ produced by *MinMix* and (b) target DMFB platform.

and the other droplet is stored and subsequently mixed with a new droplet of R_1, which is dispensed from an on-chip input reservoir. After executing the desired sequence of mix-split steps corresponding to the mixing tree, two droplets having the target ratio $\{96 : 80 : 64 : 16\}$ are generated. Note that the mixing ratio $\{96 : 80 : 64 : 16\}$ requires only five mix-split steps and six reagent droplets, i.e., the number of mix-split steps (time) and total amount of input reagents (cost) in the ratio $\{96 : 80 : 64 : 16\}$ are significantly less than those required for the ratio $\{95 : 87 : 59 : 15\}$.

Problem definition:

Given a mixture of k reagents $\mathcal{M} = \{\langle R_1, [c_{1,min}, c_{1,max}]\rangle, \langle R_2, [c_{2,min}, c_{2,max}]\rangle, \ldots, \langle R_k, [c_{k,min}, c_{k,max}]\}$, we need to choose a ratio $\{R_1 : R_2 : \cdots : R_k = c_1 : c_2 : \cdots : c_k\}$ such that the the following conditions are satisfied:

1. $c_{i,min} \leq c_i \leq c_{i,max}$, for $i = 1, 2, \ldots, k$
2. $\sum_{i=1}^{k} c_i = 1$ (validity condition)

The objective is to minimize the number of mix-split steps and/or the total number of input droplets required for generating the ratio $\{c_1 : c_2 : \cdots : c_k\}$ using *MinMix*.

Note that the first condition honors the allowable concentration range for each of constituent reagents in a mixture, whereas the second condition ensures a valid ratio.

5.3 Proposed Method

All state-of-the-art sample preparation algorithms start with an approximated target ratio in the form $\{R_1 : R_2 : \cdots : R_k = r_1 : r_2 : \cdots : r_k\}$, where $\sum_{i=1}^{k} r_i = 2^d$ and $r_i, d \in \mathbb{N}$. Note that d denotes the maximum depth of the mixing tree for $\{r_1 : r_2 : \cdots : r_k\}$. Also, for all practical purposes, $d \leq 10$ [MRB+14]. In the ratio selection method, we normalize the allowable concentration range of input reagents for a chosen value of d as follows. Given $\mathcal{M} = \{\langle R_1, [c_{1,min}, c_{1,max}]\rangle, \langle R_2, [c_{2,min}, c_{2,max}]\rangle, \ldots, \langle R_k, [c_{k,min}, c_{k,max}]\rangle\}$, and $d \in \mathbb{N}$, we normalize $\mathcal{M}$ as $\mathcal{M}' = \{\langle R_1, [\frac{\lceil 2^d \cdot c_{1,min}\rceil}{2^d}, \frac{\lfloor 2^d \cdot c_{1,max}\rfloor}{2^d}]\rangle, \langle R_2, [\frac{\lceil 2^d \cdot c_{2,min}\rceil}{2^d}, \frac{\lfloor 2^d \cdot c_{2,max}\rfloor}{2^d}]\rangle, \ldots, \langle R_k, [\frac{\lceil 2^d \cdot c_{k,min}\rceil}{2^d}, \frac{\lfloor 2^d \cdot c_{k,max}\rfloor}{2^d}]\rangle\}$. For simplicity, we can ignore the denominator in $\mathcal{M}'$ and represent $\mathcal{M}'$ as $\mathcal{M}'' = \{\langle R_1, [r_{1,min}, r_{1,max}]\rangle, \langle R_2, [r_{2,min}, r_{2,max}]\rangle, \ldots, \langle R_k, [r_{k,min}, r_{k,max}]\rangle\}$, where $r_{i,min} = \lceil 2^d \cdot c_{i,min}\rceil$ and $r_{i,max} = \lfloor 2^d \cdot c_{i,max}\rfloor$, for $i = 1, 2, \ldots k$. In this setting, our objective is to choose a target ratio $\{R_1 : R_2 : \cdots : R_k = r_1 : r_2 : \cdots : r_k\}, r_i \in \mathbb{N}$ from $\mathcal{M}''$, such that mixing time or cost is minimized subject to:

1. $r_{i,min} \leq r_i \leq r_{i,max}$, for $i = 1, 2, \ldots, k$
2. $\sum_{i=1}^{k} r_i = 2^d, d \in \mathbb{N}$ (validity condition)

The number of mix-split steps (or total reagent consumption) is determined by the number of ones in the binary representation of the normalized mixing ratio. Here, we describe a greedy heuristic approach for minimizing runs of consecutive ones in the binary representation of *CF*s of each reagent in the mixing ratio. The procedure is analogous to Booth's multiplication algorithm [Boo51]; we replace a run of consecutive ones in a valid mixing ratio for transforming it to another valid ratio that have fewer number nonzero bits. Let us consider an illustrative example for describing the transformation procedure. As described in Section 5.2, $\mathcal{M} = \{\langle R_1, [0.3, 0.4]\rangle, \langle R_2, [0.25, 0.35]\rangle, \langle R_3, [0.20, 0.05]\rangle, \langle R_4, [0.05, 0.10]\rangle\}$ is normalized to $\mathcal{M}' = \{\langle R_1, [77, 102]\rangle, \langle R_2, [64, 89]\rangle, \langle R_3, [52, 64]\rangle, \langle R_4, [13, 25]\rangle\}$ for $d = 8$. Let us choose an initial valid ratio $\{R_1 : R_2 : R_3 : R_4 = 95 : 87 : 59 : 15\}$ for which mixing tree is shown in Figure 5.3. We replace the longest run of ones in 95, i.e., $95 : 01\underline{011111}$ is replaced with $96 : 01100000$. In order to satisfy the validity conditions, we need to decrease one from another ratio for maintaining the total sum to 256. Hence, we replace $59 : 0011101\underline{1}$ with $58 : 00111010$. The total number of ones in the transformed ratio $\{R_1 : R_2 : R_3 : R_4 = \underline{96} : 87 : \underline{58} : 15\}$ is thus

(a)		(b)		(c)		(d)		(e)
95 : 01011111	$\xrightarrow{+1}$	(96) : 01100000		96 : 01100000		96 : 01100000		96 : 01100000
87 : 01010111		87 : 01010111		87 : 01010111	$\xrightarrow{+1}$	(88) : 01011000	$\xrightarrow{-8}$	(80) : 01010000
59 : 00111011	$\xrightarrow{-1}$	(58) : 00111010	$\xrightarrow{-1}$	(57) : 00111001	$\xrightarrow{-1}$	(56) : 00111000	$\xrightarrow{+8}$	(64) : 01000000
15 : 00001111		15 : 00001111	$\xrightarrow{+1}$	(16) : 00010000		16 : 00010000		16 : 00010000

Figure 5.5 Ratio transformation for $\mathcal{M}' = \{\langle R_1, [77, 102]\rangle, \langle R_2, [64, 89]\rangle, \langle R_3, [52, 64]\rangle, \langle R_4, [13, 25]\rangle\}$.

decreased by 5. Figure 5.5(a)–(b) show this transformation step. In the next transformation step (Figure 5.5(b)–(c)), $15 : 00001111$ and $58 : 00111010$ are replaced with $16 : 00010000$ and $57 : 00111001$, leading to another valid mixing ratio $\{R_1 : R_2 : R_3 : R_4 = 96 : 87 : \underline{57} : \underline{16}\}$. After two more transformation steps, the ratio becomes $\{R_1 : R_2 : R_3 : R_4 = 96 : 80 : 64 : 16\}$, for which no further transformation is possible. The complete procedure for ratio transformation is shown in Figure. 5.5(a)–(e).

Although this heuristic procedure for ratio transformation procedure reduces the number of ones in the final ratio, it is highly dependent on the initial ratio and the choice of two *CF*s participating in each transformation step. Hence, it may not guarantee the optimality of the final outcome. Moreover, if the cost of each reagent varies, then "minimizing the longest run of ones first" heuristic does not reduce the total reagent cost. We need to consider the cost of input reagents in the ratio transformation procedure, thereby complicating the ratio selection process even more. In the following subsection, we discuss an integer linear programming (ILP) formulation of the problem, which captures the above-mentioned ratio-transformation process considering both uniform and weighted costs of reagents and produces an optimal solution.

5.3.1 An ILP Formulation for Optimal Solution

Assume that we are given $\mathcal{M} = \{\langle R_1, [r_{1,min}, r_{1,max}]\rangle, \langle R_2, [r_{2,min}, r_{2,max}]\rangle, \ldots, \langle R_k, [r_{k,min}, r_{k,max}]\rangle\}, d \in \mathbb{N}$, and the cost of each input reagents $c(R_i) = w_i \geq 1$, for $i = 1, 2, \ldots, k$, where $c(R_i)$ denotes the cost of reagent R_i. Let the d-bit binary representation of *CF* (R_i) be $b^i_{d-1}, b^i_{d-2}, \ldots, b^i_0$, for $i = 1, 2, \ldots, k$. We need to define a Boolean variable for each bit in the d-bit binary representation of *CF* of R_i. Hence, the set of Boolean variables is $\{b^i_{d-1}, b^i_{d-2}, \ldots, b^i_0$, for $i = 1, 2, \ldots, k\}$. Note that the number of non-zero bits in the binary representation of *CF* determines the required number of input droplets for that sample. Hence, we need to consider the cost of each input reagent when minimizing the number of ones in the target ratio. The objective function is defined as follows.

$$\text{Minimize: } \sum_{i=1}^{k}\sum_{j=0}^{d-1} w_i \cdot b_j^i \tag{5.1}$$

subject to:

$$r_{i,min} \le \sum_{j=0}^{d-1} 2^j \cdot b_j^i \le r_{i,max} \qquad \text{for } i = 1, 2, \ldots, k$$

$$\sum_{i=1}^{k}\sum_{j=0}^{d-1} 2^j \cdot b_j^i = 2^d \tag{5.2}$$

Note that the number of variables and constraints in the ILP instance of $\mathcal{M}$ are $d \cdot k$ and $2 \cdot k + 1$, respectively. Since, for all practical purposes, $d \le 10$, the number of variables and constraints in the ILP will be $O(k)$, i.e., linear in the number of input reagents in the mixture. Moreover, if we choose $w_1 = w_2 = \cdots = w_k = 1$, the ILP minimizes the number of mix-split steps in the mixing tree generated by *MinMix*.

Let us consider an illustrative example for $\mathcal{M} = \{\langle R_1, [77, 102]\rangle, \langle R_2, [64, 89]\rangle, \langle R_3, [52, 64]\rangle, \langle R_4, [13, 25]\rangle\}$, $d = 8$ and $w_1 = w_2 = w_3 = w_4 = 1$. The Boolean variables are $b_7^i, b_6^i, \ldots, b_0^i$, for $i = 1, 2, \ldots, k$. The ILP formulation is given below.

$$\text{Minimize: } \sum_{i=1}^{4}(b_7^i + b_6^i + \cdots + b_0^i)$$

Subject to:

$$77 \le 2^7 \cdot b_7^1 + 2^6 \cdot b_6^1 + \cdots + 2^0 \cdot b_0^1 \le 102$$
$$64 \le 2^7 \cdot b_7^2 + 2^6 \cdot b_6^2 + \cdots + 2^0 \cdot b_0^2 \le 89$$
$$52 \le 2^7 \cdot b_7^3 + 2^6 \cdot b_6^3 + \cdots + 2^0 \cdot b_0^3 \le 64$$
$$13 \le 2^7 \cdot b_7^4 + 2^6 \cdot b_6^4 + \cdots + 2^0 \cdot b_0^4 \le 25$$
$$\sum_{i=1}^{4}(2^7 \cdot b_7^i + 2^6 \cdot b_6^i + \cdots + 2^0 \cdot b_0^i) = 2^8$$

The solution of ILP produces the following target ratio $\{R_1 : R_2 : R_3 : R_4 = 96 : 80 : 64 : 16\}$. The corresponding mixing tree is shown in Figure 5.4.

5.4 Experimental Results

We evaluate the performance of the ILP-based technique for choosing a target ratio from a set of valid target ratios that minimizes the sample preparation time or cost. For ILP-solver, we use CPLEX [cpl] optimizer. We consider several mixture (Table 5.1), where the *CF* of each reagent lies within an allowable range. Depending on the input mixture specification and the maximum depth of the mixing tree (d), we report the number of mix-split steps (mixture-preparation time) and total reagent requirement (cost) in the mixing tree generated by *MinMix* for the chosen target ratio. Moreover, the depth of the mixing tree for each selected ratio (d') is also shown. In these experiments, we assume the uniform cost metric for each constituent reagent in a mixture. Table 5.2 shows two ratios for each input ratio space having minimum and maximum mixing steps, and reagent consumption when *MinMix* algorithm is used. We also tabulate the number of each reagent in an array named as reagent vector. It can be observed from Table 5.2 that the difference between the number of mixing steps needed for the best and the worst ratio (for each $\mathcal{M}_i$) is quite significant. In other words, the selected ratio strongly influences sample-preparation time and cost. Note that the CPU-time for running the ILP-solver is negligible even in larger ratio space. Additionally, we consider non-uniform cost metric for input reagents in a mixture and run ILP-solver for finding optimal mixing ratio. Note that we consider the same input mixtures that are used in Table 5.1 with different cost assignments for input reagents and report the desired ratio along with several other performance parameters in Table 5.3 for *MinMix* and *CoDOS*.

Note that once a ratio is selected by the ratio selection algorithm, different strategies may be adopted to produce the final mixing tree (Figure 5.1). As

Table 5.1 Specification of input ratios

	Input Mixture
$\mathcal{M}_1$	$\{\langle R_1, [0.54, 0.70]\rangle, \langle R_2, [0.172, 0.235]\rangle, \langle R_3, [0.14, 0.18]\rangle\}$
$\mathcal{M}_2$	$\{\langle R_1, [0.3, 0.4]\rangle, \langle R_2, [0.2, 0.3]\rangle, \langle R_3, [0.22, 0.28]\rangle, \langle R_4, [0.05, 0.15]\rangle\}$
$\mathcal{M}_3$	$\{\langle R_1, [0.4, 0.5]\rangle, \langle R_2, [0.18, 0.25]\rangle, \langle R_3, [0.18, 0.25]\rangle, \langle R_4, [0.02, 0.1]\rangle, \langle R_5, [0.035, 0.050]\rangle\}$
$\mathcal{M}_4$	$\{\langle R_1, [0.25, 0.30]\rangle, \langle R_2, [0.2, 0.3]\rangle, \langle R_3, [0.15, 0.20]\rangle, \langle R_4, [0.1, 0.2]\rangle, \langle R_5, [0.05, 0.10]\rangle\},$ $\langle R_6, [0.03, 0.08]\rangle\}$
$\mathcal{M}_5$	$\{\langle R_1, [0.16, 0.20]\rangle, \langle R_2, [0.11, 0.18]\rangle, \langle R_3, [0.1, 0.3]\rangle, \langle R_4, [0.1, 0.4]\rangle, \langle R_5, [0.1, 0.2]\rangle\},$ $\langle R_6, [0.1, 0.3]\rangle, \langle R_7, [0.05, 0.10]\rangle\}$

Table 5.2 Performance of the ratio selection method considering the uniform cost for all input reagents

Input Mixture	d	Total Ratios	Ratio Selected by the Ratio Selection Method Using *MinMix*	d'	n_m (n_w)	Reagent Vector	Cost	CPU-Time
$\mathcal{M}_1$	8	176	$\{160 : 56 : 40\}$	5	6	$[2, 3, 2]$	7	17 ms.
			$\{159 : 55 : 42\}$	8	13	$[6, 5, 3]$	14	26 ms.
$\mathcal{M}_2$	8	4819	$\{96 : 64 : 64 : 32\}$	3	4	$[2, 1, 1, 1]$	5	17 ms.
			$\{95 : 63 : 63 : 35\}$	8	20	$[6, 6, 6, 3]$	21	20 ms.
$\mathcal{M}_3$	8	20632	$\{128 : 64 : 48 : 6 : 10\}$	7	7	$[1, 1, 2, 2, 2]$	8	66 ms.
			$\{111 : 63 : 63 : 7 : 12\}$	8	22	$[6, 6, 6, 3, 2]$	23	28 ms.
$\mathcal{M}_4$	8	512805	$\{64 : 64 : 48 : 48 : 16 : 16\}$	4	7	$[1, 1, 2, 2, 1, 1]$	8	24 ms.
			$\{69 : 73 : 47 : 47 : 15 : 15\}$	8	26	$[3, 6, 5, 5, 4, 4]$	27	38 ms.
$\mathcal{M}_5$	8	52911861	$\{48 : 32 : 32 : 32 : 32 : 64 : 16\}$	4	7	$[2, 1, 1, 1, 1, 1, 1]$	8	21 ms.
			$\{47 : 31 : 31 : 31 : 30 : 63 : 23\}$	8	33	$[5, 5, 5, 5, 4, 6, 4]$	34	33 ms.

d is the maximum depth of the mixing tree given as input; d' is the actual depth of the mixing tree (generated by *MinMix*) for the ratio selected by the ratio selection method; $n_m = n_w$ since *MinMix* algorithm is used.

Table 5.3 Performance of the ratio selection method considering the non-uniform cost for all input reagents

Input	d	Cost Vector	Ratio Selected by the Ratio Selection Method	d'	*MinMix*				*CoDOS* Run on the Ratio Selected by *MinMix*			
					$n_m(n_w)$*	Reagent Vector	Cost	CPU-Time	n_m	n_w	Reagent Vector	Cost
$\mathcal{M}_1$	8	[1, 1, 1]	{160 : 56 : 40}	5	6	[2, 3, 2]	7	17 ms.	5	4	[2, 2, 1]	5
		[1, 2, 1]	{168 : 48 : 40}	5	6	[3, 2, 2]	9	17 ms.	5	4	[2, 2, 1]	7
		[5, 2, 1]	{160 : 56 : 40}	5	6	[2, 3, 2]	18	16 ms.	5	4	[2, 2, 1]	15
$\mathcal{M}_2$	8	[1, 1, 1, 1]	{96 : 64 : 64 : 32}	3	4	[2, 1, 1, 1]	5	17 ms.	4	4	[2, 1, 1, 1]	5
		[5, 1, 1, 1]	{96 : 64 : 64 : 32}	3	4	[2, 1, 1, 1]	13	17 ms.	4	4	[2, 1, 1, 1]	13
		[1, 5, 1, 1]	{96 : 64 : 64 : 32}	3	4	[2, 1, 1, 1]	9	18 ms.	4	4	[2, 1, 1, 1]	9
$\mathcal{M}_3$	8	[1, 1, 1, 1, 1]	{128 : 64 : 48 : 6 : 10}	7	7	[1, 1, 2, 2, 2]	8	66 ms.	7	6	[1, 1, 2, 1, 2]	7
		[1, 1, 5, 5, 1]	{108 : 64 : 64 : 8 : 12}	6	8	[4, 1, 1, 1, 2]	17	27 ms.	7	6	[3, 1, 1, 1, 1]	15
		[5, 1, 5, 5, 1]	{128 : 47 : 64 : 8 : 9}	8	9	[1, 5, 1, 1, 2]	22	44 ms.	8	7	[1, 4, 1, 1, 1]	20
$\mathcal{M}_4$	8	[1, 1, 1, 1, 1, 1]	{64 : 64 : 48 : 48 : 16 : 16}	4	7	[1, 1, 2, 2, 1, 1]	8	24 ms.	6	5	[1, 1, 1, 1, 1, 1]	6
		[1, 1, 1, 5, 5, 1]	{72 : 72 : 48 : 32 : 16 : 16}	5	8	[2, 2, 2, 1, 1, 1]	17	24 ms.	7	6	[1, 1, 2, 1, 1, 1]	15
		[5, 1, 1, 5, 5, 1]	{64 : 76 : 48 : 32 : 16 : 20}	6	9	[1, 3, 2, 1, 1, 2]	22	19 ms.	6	5	[1, 2, 2, 1, 1, 2]	21
$\mathcal{M}_5$	8	[1, 1, 1, 1, 1, 1, 1]	{48 : 32 : 32 : 32 : 32 : 64 : 16}	4	7	[2, 1, 1, 1, 1, 1, 1]	8	21 ms.	8	7	[1, 1, 1, 1, 1, 2, 1]	8
		[5, 1, 1, 1, 1, 1, 1]	{48 : 32 : 32 : 32 : 32 : 64 : 16}	4	7	[2, 1, 1, 1, 1, 1, 1]	16	27 ms.	8	7	[1, 1, 1, 1, 1, 2, 1]	12
		[5, 1, 1, 5, 5, 1, 1]	{48 : 32 : 32 : 32 : 32 : 64 : 16}	4	7	[2, 1, 1, 1, 1, 1, 1]	24	25 ms.	8	7	[1, 1, 1, 1, 1, 2, 1]	20

* $n_m = n_w$ for *MinMix*.

discussed in Section 5.1, *CoDOS* starts with *MinMix*-tree and systematically transforms it into a reduced mixing graph. The new ratio selection method, on the other hand, utilizes the underlying combinatorial properties of the*MinMix* algorithm to determine a ratio for which the mixing tree has minimum number of mix-split operations; *CoDOS* is then invoked only once on this ratio to further optimize reactant cost, as opposed to multiple calls made to *CoDOS* routine by *Deputy* and *Fusion* algorithms. Moreover, the ratio selected by the ratio selection method can be used to run other sample-preparation algorithms such as *RMA* [RCK+15b] and *MTCS* [KRC+13]. The last four columns of Table 5.3 show different parameters when the selected ratios are run with *CoDOS*. It can be observed from Table 5.3 and Table 5.4 that for $\mathcal{M}_1$, $\mathcal{M}_2$, and $\mathcal{M}_5$, the costs for the desired target ratios (using *CoDOS*) returned by the new ratio selection method are identical to those returned by *Deputy* and *Fusion*. For $\mathcal{M}_3$ and $\mathcal{M}_4$, *Deputy* (*Fusion*) performs marginally better for three (one) cases out of six cases. However, the *CoDOS* calls needed for finding such ratios using *Deputy* and *Fusion* are quite large in number.

We run three different sample-preparation algorithms *RMA* [RCK+15b], *MTCS* [KRC+13], and *CoDOS* [LCLH13] exhaustively for the entire space of $\mathcal{M}_1, \mathcal{M}_2$, and $\mathcal{M}_3$ (Table 5.1 and Table 5.2), and report the one having the minimum number of mixing operations (marked with †) in Table 5.5. We also record the best ratio found by the ratio selection method (marked with $*$) and compute the performance parameters (number of mixing steps (n_m), waste droplets (n_w), and reagent vector) of the selected ratio for *MinMix* [TUTA08]. A ratio marked with †$*$ in Table 5.5 represents the fact that the ratio selected by the new ratio selection method is same as the best ratio obtained by running *RMA/MTCS/CoDOS* exhaustively over the entire valid input-ratio space. In case of $\mathcal{M}_1$, the ratio selected by the ILP instance also produces minimum mixing using *MTCS* and *CoDOS*. However, in the case of *RMA*, a different ratio yields minimum number of mixing. We also run the ratio returned by ratio selection method using *RMA* and observe that it requires only one additional mixing step compared to the ratio that requires minimum mix-split steps for $\mathcal{M}_1$ when *RMA* is invoked. The ratio selected by ILP gives minimum number of mixing for *RMA*, *MTCS*, and *CoDOS* in the case of $\mathcal{M}_2$. However, in the case of $\mathcal{M}_3$, the selected ratio gives minimum mixing in *RMA*, but it takes one extra mixing step for *MTCS* and *CoDOS*. For ratio space $\mathcal{M}_4$ and $\mathcal{M}_5$ (last two rows of Table 5.5), we could not run all possible ratios using *RMA* and *MTCS* as the space is very large. Instead, we consider the ratios returned by ILP-solver when run on *RMA*, *MTCS*, and *CoDOS*

Table 5.4 Results of *Deputy* and *Fusion* algorithms [LCg16] using *CoDOS*

Mixture	Cost Vector	*Deputy*						*Fusion*					
		Ratio	n_m	n_w	Reagent Vector	Cost	#Calls[a]	Ratio	n_m	n_w	Reagent Vector	Cost	#Calls[a]
$\mathcal{M}_1$	[1, 1, 1]	{160, 56, 40}	5	4	[2, 2, 1]	5	37	{160, 56, 40}	5	4	[2, 2, 1]	5	37
	[1, 2, 1]	{160, 56, 40}	5	4	[2, 2, 1]	7	27	{168, 48, 40}	5	4	[2, 2, 1]	7	27
	[5, 2, 1]	{160, 56, 40}	5	4	[2, 2, 1]	15	37	{160, 56, 40}	5	4	[2, 2, 1]	15	37
$\mathcal{M}_2$	[1, 1, 1, 1]	{88, 72, 64, 32}	5	4	[2, 1, 1, 1]	5	58	{96, 64, 64, 32}	4	4	[2, 1, 1, 1]	5	65
	[5, 1, 1, 1]	{88, 72, 64, 32}	5	4	[2, 1, 1, 1]	13	58	{96, 64, 64, 32}	4	4	[2, 1, 1, 1]	13	65
	[1, 5, 1, 1]	{96, 64, 64, 32}	4	4	[2, 1, 1, 1]	9	59	{96, 64, 64, 32}	4	4	[2, 1, 1, 1]	9	59
$\mathcal{M}_3$	[1, 1, 1, 1, 1]	{104, 52, 64, 24, 12}	8	6	[2, 2, 1, 1, 1]	7	44	{128, 64, 48, 6, 10}	7	6	[1, 1, 2, 1, 2]	7	49
	[1, 1, 5, 5, 1]	{108, 60, 60, 16, 12}	8	5	[2, 1, 1, 1, 1]	14	59	{104, 60, 64, 16, 12}	9	7	[2, 3, 1, 1, 1]	16	60
	[5, 1, 5, 5, 1]	{128, 56, 48, 12, 12}	7	5	[1, 2, 1, 1, 1]	18	55	{128, 47, 64, 8, 9}	8	7	[1, 4, 1, 1, 1]	20	48
$\mathcal{M}_4$	[1, 1, 1, 1, 1, 1]	{64, 64, 48, 48, 16, 16}	6	5	[1, 1, 1, 1, 1, 1]	6	80	{64, 64, 48, 48, 16, 16}	6	5	[1, 1, 1, 1, 1, 1]	6	80
	[1, 1, 1, 5, 5, 1]	{72, 54, 48, 48, 16, 18}	8	6	[1, 2, 1, 1, 1, 1]	15	80	{64, 76, 51, 32, 16, 17}	8	7	[1, 2, 2, 1, 1, 1]	16	60
	[5, 1, 1, 5, 5, 1]	{72, 54, 48, 48, 16, 18}	8	6	[1, 2, 1, 1, 1, 1]	19	80	{64, 76, 51, 32, 16, 17}	8	7	[1, 2, 2, 1, 1, 1]	20	60
$\mathcal{M}_5$	[1, 1, 1, 1, 1, 1, 1]	{48, 32, 32, 32, 32, 56, 24}	8	7	[1, 1, 1, 1, 1, 2, 1]	8	173	{48, 32, 48, 32, 32, 40, 24}	9	7	[1, 1, 1, 1, 1, 1]	8	157
	[5, 1, 1, 1, 1, 1, 1]	{48, 32, 32, 32, 32, 56, 24}	8	7	[1, 1, 1, 1, 1, 2, 1]	12	173	{48, 32, 48, 32, 32, 40, 24}	9	7	[1, 1, 1, 1, 1, 1]	12	157
	[5, 1, 1, 5, 5, 1, 1]	{48, 32, 32, 32, 32, 56, 24}	8	7	[1, 1, 1, 1, 1, 2, 1]	20	179	{48, 32, 48, 32, 32, 40, 24}	9	7	[1, 1, 1, 1, 1, 1]	20	163

[a] *number of calls to CoDOS.*

Table 5.5 Comparative results for *RMA* [RCK+15b], *MTCS* [KRC+13], and *CoDOS* [LCLH13]

Mixture	Method	Ratio	n_m	n_w	Reagent Vector	Cost
$\mathcal{M}_1$	*RMA*	$\{168, 48, 40\}^{\dagger}$	7	7	$[4, 2, 2]$	8
		$\{160, 56, 40\}^{*}$	8	8	$[4, 3, 2]$	9
	MTCS	$\{160, 56, 40\}^{\dagger *}$	5	4	$[2, 2, 1]$	5
	CoDOS	$\{160, 56, 40\}^{\dagger *}$	5	4	$[2, 2, 1]$	5
$\mathcal{M}_2$	All	$\{94 : 64 : 64 : 32\}^{\dagger *}$	4	4	$[2, 1, 1, 1]$	5
$\mathcal{M}_3$	*RMA*	$\{128 : 64 : 48 : 6 : 10\}^{\dagger *}$	7	7	$[1, 1, 2, 2, 2]$	8
	MTCS	$\{128 : 48 : 56 : 12 : 12\}^{\dagger}$	6	5	$[1, 1, 2, 1, 1]$	6
		$\{128 : 64 : 48 : 6 : 10\}^{*}$	7	7	$[1, 1, 2, 2, 2]$	8
	CoDOS	$\{108 : 48 : 64 : 24 : 12\}^{\dagger}$	6	5	$[2, 1, 1, 1, 1]$	6
		$\{128 : 64 : 48 : 6 : 10\}^{*}$	7	6	$[1, 1, 2, 1, 2]$	7
$\mathcal{M}_4$‡	*RMA*	$\{64, 64, 48, 48, 16, 16\}^{*}$	7	7	$[1, 1, 2, 2, 1, 1]$	8
	MTCS		6	5	$[1, 1, 1, 1, 1, 1]$	6
	CoDOS		6	5	$[1, 1, 1, 1, 1, 1]$	6
$\mathcal{M}_5$‡	*RMA*	$\{48, 32, 32, 32, 32, 64, 16\}^{*}$	7	7	$[2, 1, 1, 1, 1, 1, 1]$	8
	MTCS		7	7	$[2, 1, 1, 1, 1, 1, 1]$	8
	CoDOS		8	7	$[1, 1, 1, 1, 1, 2, 1]$	8

*ratio returned by the ratio selection method (optimal for *MinMix*).

†best ratio found by running the corresponding sample-preparation algorithm exhaustively over the entire valid ratio-space.

‡unable to run the corresponding sample-preparation algorithm exhaustively as the valid ratio-space is too large.

and report performance parameters. It is quite evident from experimental results that choosing an optimal ratio using *MinMix* indeed performs well for *RMA*, *MTCS*, and *CoDOS* also.

We have also considered few real-life mixing ratios, taken from [ASS+17], and run several experiments by considering variability of input reagents in the mixing ratios. Table 5.6 shows the desired target ratio along with several other parameters by considering uniform cost for every reagent.

Table 5.6 Results for real-life test cases using the new ratio selection technique assuming uniform cost for all reagents

Real-Life Bioprotocols (Ratio Used)	Approximated Ratio ($d = 8$)	Ratio Space (4% Variability)	New Ratio Selection Method	
			Selected Ratio	$n_m(n_w)$*
Plasmid DNA [15] {20:10:2:2:2:1:53}	{57:28:6:6:6:3:150}	{[47,67] : [18,38] : [1,16] : [1,16] : [1,16] : [1,13] : [140,160]}	{64:32:8:2:2:4:144}	8
Splinkerette PCR [UMK+09] {40:10:1:1:48}	{102:26:3:3:122}	{[92,112] : [16,36] : [1,13] : [1,13] : [112,132]}	{96:16:8:8:128}	6
Touchdown PCR [17] {20:12:40:20:4:4:1:99}	{26:15:51:26:5:5:1:127}	{[16,36] : [5,25] : [41,61] : [16,36] : [1,15] : [1,15] : [1,11] : [117,137]}	{32:16:48:16:4:4:8:128}	9
Silver-Restriction Digest [17] {70:10:10:2:2:6}	{180:26:26:5:5:14}	{[170,190] : [16,36] : [16,36] : [1,15] : [1,15] : [4,24]}	{176:32:16:8:8:16}	8
Molecular barcodes - PCR [17] {50:10:10:10:10:25:25:50:2:308}	{26:5:5:5:5:13:13:26:1:157}	{[16,36] : [1,15] : [1,15] : [1,15] : [1,15] : [3,23] : [3,23] : [16,36] : [1,11] : [147,167]}	{32:1:1:1:1:4:16:32:8:160}	11

*$n_n = n_w$ *for MinMix.*

5.5 Conclusions

We have explained an efficient ratio-selection technique for sample-preparation based on *MinMix* algorithm [TUTA08], by exploiting the inherent variability in ratio-specification. Among the entire permissible ratio-space, our method selects the one that requires the minimum reactant-cost when *MinMix* algorithm is used to generate the mixing tree. Moreover, the selected ratio performs favorably when other state-of-the-art sample preparation algorithms such as *RMA* [RCK+15b], *MTCS* [KRC+13], and *CoDOS* [LCLH13] are invoked on the ratio selected by the new ratio selection algorithm. Hence, this technique can be viewed as a general-purpose pre-processing wrapper for several sample-preparation algorithms for the purpose of saving sample-preparation time and cost significantly when concentration-resilience in the target mixture is allowed. Formulation of the underlying optimization problem for selecting a suitable ratio that minimizes reactant-cost for *RMA*-, *MTCS*-, or *CoDOS*-based algorithms remains open to settle.

6

Dilution and Mixing Algorithms for Flow-based Microfluidic Biochips

Albeit sample preparation is well studied for digital microfluidic biochips, very few prior work addressed this problem in the context of continuous-flow microfluidics from an algorithmic perspective. In the continuous flow microfluidic biochips (CFMBs), microvalves, and micropumps are used to manipulate on-chip fluid flow through microchannels in order to execute a biochemical protocol. The detailed description of CFMBs and their fluidic operations are elaborated in Section 1.1. CFMBs are quite popular in the biochemistry community because of their simplicity of fabrication, flexibility in reagent-volume control, and versatility of applications such as automation of assays and point-of-care diagnosis [CLS12]. Many clinical applications, e.g., DNA analysis [DY11], drug discovery [NGL+12], and sample preparation [SQH+15], have been successfully implemented on CFMB. In most of the biochemical applications, automatic sample preparation plays a significant role, e.g., it accounts for 60% of the work in analytical tests [DZN04]. Microfluidic networks based on CFMB can be used to produce logarithmic and linear concentration gradients effectively [KLK+08, NTFF04]. Moreover, flow-based microfluidic devices have been developed for implementing serial and parallel mixing, and also for producing dilution gradients and diffusion-based gradients [NTFF04].

In recent years, many sample preparation algorithms [HHC14, TUTA08, RBC10, RBC11, LCH15, CLH13, HLL13, RCK+15b, LCLH13, KRC+13, DYHH13] have been proposed for preparing a mixture of fluids with a digital microfluidic biochip (DMFB), based on the (1:1) mixing model. On a continuous flow-based microfluidic platform, an N-segment rotary mixer that is divided into N equal-length segments (henceforth, denoted as Mixer-N), can be conveniently used to implement more powerful and multiple mixing models [HLH15]. Rotary mixers with micropumps use circular motion to mix two fluids that are pumped into a ring-like or rotary structure [MQ07]. However, all previous works [LSH15, HLH15] fail to exploit the full functionality of different mixing models that are supported by a rotary mixer as they impose several constraints on allowable mixing steps. It is also observed that flow-based microfluidic mixing operations are quite slow [LCWF11b], and hence, it is desirable to reduce such operations as far as possible during sample preparation.

In this chapter, a dilution algorithm for CFMBs is presented that can be used to prepare a mixture of two reagents (sample and buffer) using a rotary mixer that minimizes the number of mixing operations and reduces reagent usage. Second, a generalized approach is discussed for a reagent-saving mixture preparation algorithm, where the number of reagents may be greater than two.

A novel formulation of the dilution problem is presented using a set of linear constraints over integers, and then a solver based on *Satisfiability Modulo Theory* (SMT) [dMB08] is invoked to obtain the optimal solution. The proposed algorithm is guaranteed to produce a sample of desired target concentration factor within a tolerable error limit, while minimizing the number of mixing operations as a "primary objective", and minimizing reagent usage as a "secondary objective" criterion. Unlike existing methods [LSH15, HLH15], a two-step optimization technique is deployed: first, a mixing tree that consists of a minimum number of mixing steps is constructed; next, using an SMT-based approach, the proposed technique searches among all such trees to find the one that produces the desired target concentration with minimum sample usage. Simulation results show that for an eight-segment mixer, the proposed dilution algorithm terminates very fast (in around 0.34 second on a 2 GHz, Intel Core i5 processor), and on the average, it reduces the number of mixing operations and waste production compared to prior work *TPG* [LSH15] and *VOSPA* [HLH15]. Furthermore, with respect to reagent usage, it improves upon *TPG* and compares favorably with *VOSPA*. Moreover, the performance of mixture-preparation algorithms is also studied in detail.

6.1 Sample Preparation and Mixing Models

In order to ensure the self-containment of this chapter, basic sample preparation and mixing models are revisited once again. Sample preparation is the process of mixing two or more biochemical fluidic reagents in a given volumetric ratio through a sequence of mixing operations. Dilution is a special case of sample preparation, where only two input reagents (commonly known as sample and buffer) are mixed in a desired volumetric ratio. In a DMFB, two equal-volume droplets are mixed using the (1:1) mixing model [HHC14, TUTA08, RBC10, RBC11, LCH15, CLH13, HLL13, RCK+15b, LCLH13, KRC+13, DYHH13]. On the other hand, a CFMB supports multiple ratios as mixing models by deploying a Mixer-N [HLH15]. A Mixer-N is divided into N equal-length segments, where each segment can be filled with a fluid.

Example 6.1.1 Figure 6.1 shows a Mixer-4, and various mixing models supported by this device. Different valves are used for loading and unloading a segment (gate valve), for determining the segment boundary (ratio valve), and for mixing of fluids within the segments (pumping valve). A description of such mixers can be found in Section 1.1. Figures 6.1(b)–6.1(e) show each of the four different reagents with a different color. ■

A mixture of k reagents $R_1, R_2, \ldots, R_k$ is denoted as $\mathcal{M} = \{\langle R_1, c_1\rangle, \langle R_2, c_2\rangle, \ldots, \langle R_k, c_k\rangle\}$, where $\sum_{i=1}^{k} c_i = 1$ (validity condition) and $0 \leq c_i \leq 1$ for $i = 1, 2, \ldots, k$. In other words, $R_1, R_2, \ldots, R_i, \ldots, R_k$ are mixed with a ratio of $\{c_1 : c_2 : \cdots : c_i : \cdots : c_k\}$, where c_i denotes the concentration factor (*CF*) of R_i. The condition $\sum_{i=1}^{k} c_i = 1$ [TUTA08] ensures the validity of a mixing ratio. Note that the *CF* of a pure (100%) reagent R_i is assumed to be 1, and the *CF* of neutral buffer (0%) is assumed to be 0. Because of the inherent mixing models used for microfluidic implementation of mixers, each c_i is required to be approximated as a special form

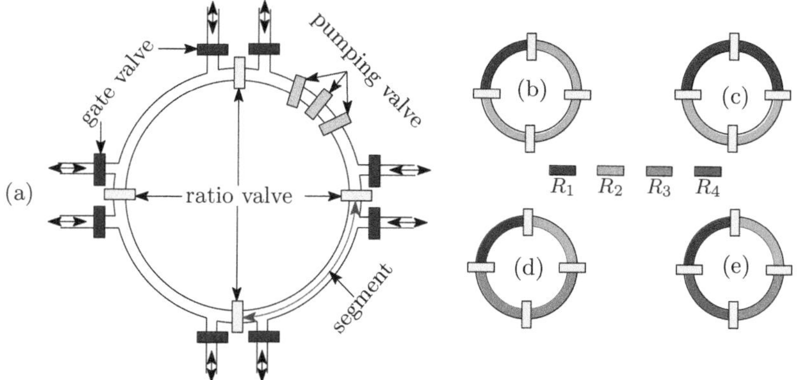

Figure 6.1 (a) A four-segment rotary mixer (Mixer-4) and its possible mixing models (b) $(1:3)$, (c) $(1:1)$, (d) $(1:1:2)$, and (e) $(1:1:1:1)$.

$\frac{x_i}{N^d}$, where x_i, N, and d are positive integers, i.e., $x_i, N, d \in \mathbb{N}$. The value of N is determined by the mixing model supported by the microfluidic mixer, and d is determined by the desired accuracy of approximation (error-tolerance limit ϵ in *CF*, $0 \leq \epsilon < 1$), which is user-specified. A lower value of ϵ denotes higher accuracy in *CF*. Assuming the underlying mixing model of Mixer-N and for a given value of error-tolerance limit ϵ, we choose the minimum value of d such that each c_i in mixture $\mathcal{M}$ is approximated as $\frac{x_i}{N^d}$, where $x_i \in \mathbb{N}$ and $\max_i\{|c_i - \frac{x_i}{N^d}|\} < \epsilon$ and $\sum_{i=1}^{k} \frac{x_i}{N^d} = 1$. In other words, each c_i is approximated as a d-digit N-ary fraction. For example, in case of (1:1) mixing model ($N = 2$) and a given ϵ, mixture $\mathcal{M}$ needs to be approximated as $\{\frac{y_1}{2^d} : \frac{y_2}{2^d} : \cdots : \frac{y_k}{2^d}\} \equiv \{y_1 : y_2 : \cdots : y_k\}$ by choosing the smallest d such that $\max_i\{|c_i - \frac{y_i}{2^d}|\} < \epsilon$ and $\sum_{i=1}^{k} y_i = 2^d$. The following example illustrates the ratio approximation procedure and the choice of d.

Example 6.1.2 Consider a mixture $\mathcal{M} = \{\langle R_1, c_1 = 0.25\rangle, \langle R_2, c_2 = 0.30\rangle, \langle R_3, c_3 = 0.45\rangle\}$, which is to be generated using $(1:1)$ model with an error-tolerance limit $\epsilon = 0.004$. Table 6.1 shows the detailed approximation procedure; the value of d has been increased iteratively starting from 1 to find a solution. The rounding procedure for $d = 4$ is shown in the footnote of Table 6.1. Note that for $d \leq 5$, the error-tolerance limit is not satisfied, and for $d = 6$, the approximated ratio is $\mathcal{M}' = \{\langle R_1, c_1' = \frac{16}{2^6}\rangle, \langle R_2, c_2' = \frac{19}{2^6}\rangle, \langle R_3, c_3' = \frac{29}{2^6}\rangle\}$ or $\{R_1 : R_2 : R_3 = 16 : 19 : 29\}$ that satisfies the given error-tolerance and the validity condition: $16 + 19 + 29 = 2^6$. ■

A mixing graph [RBC10, LCH15] is commonly used to represent the dependencies of mixing operations, where a node with zero in-degree (out-degree) represents an input reagent (target ratio). The internal nodes represent intermediate mixing operations. In the domain of *CFMB*, an incoming

Table 6.1 Approximating a target ratio based on mixing model and user-defined error-tolerance in *CF*

d	Approx. Ratio	Error in *CF*
4	$4:5:7^{\dagger}$	$\max\{\lvert 0.25 - \frac{4}{2^4}\rvert, \lvert 0.30 - \frac{5}{2^4}\rvert, \lvert 0.45 - \frac{7}{2^4}\rvert\} = 0.01 > \epsilon$
5	$8:10:14$	$\max\{\lvert 0.25 - \frac{8}{2^5}\rvert, \lvert 0.30 - \frac{10}{2^5}\rvert, \lvert 0.45 - \frac{14}{2^5}\rvert\} = 0.01 > \epsilon$
6	$16:19:29$	$\max\{\lvert 0.25 - \frac{16}{2^6}\rvert, \lvert 0.30 - \frac{19}{2^6}\rvert, \lvert 0.45 - \frac{29}{2^6}\rvert\} = 0.003 < \epsilon$

$^{\dagger}16 \times 0.25 = 4$; $16 \times 0.3 = 4.8 \approx 5$; $16 \times 0.45 = 7.2 \approx 7$; and $4 + 5 + 7 = 2^4$.
For $d = 1, 2, 3$, $\max_{i \in \{1,2,3\}}\{|c_i - \frac{x_i}{2^d}|\} \geq \epsilon$, where c_i is approximated as $\frac{x_i}{2^d}$.

edge-weight denotes the number of segments that are to be filled with input reagents, or fluids with previously obtained intermediate *CF*s for the concerned mixing operation.

In the case of dilution problem, the *CF* of the mixture denotes the volumetric ratio of sample and buffer fluids, i.e., the portion of sample fluid present in the mixture fluid (hence, $0 \leq CF \leq 1$). Assuming that N segments of a Mixer-N are filled with *CF*s $c_1, c_2, \cdots, c_N$, of a fluid diluted with the same buffer; then, the resultant *CF* of the mixture will be $\frac{\sum_{i=1}^{N} c_i}{N}$ after homogeneous mixing. In the case of dilution, the mixing graph is commonly known as dilution graph. Figure 6.2 shows dilution graphs corresponding to the target ratio $\{\langle\text{sample}, \frac{125}{4^4}\rangle, \langle\text{buffer}, \frac{131}{4^4}\rangle\}$, or $\{\text{sample} : \text{buffer} = 125 : 131\}$ obtained by four different algorithms, when Ring-3[1] is used for *TPG* [LSH15], and Mixer-4 is used for *NWayMix*, *VOSPA* [HLH15], and *FloSPA-D*.

6.2 Related Work

The algorithmic aspects of sample preparation with a CFMB have recently been studied by many researchers. Liu *et al.* [LSH15] developed the tree pruning and grafting algorithm (*TPG*) by transforming an initial mixing graph

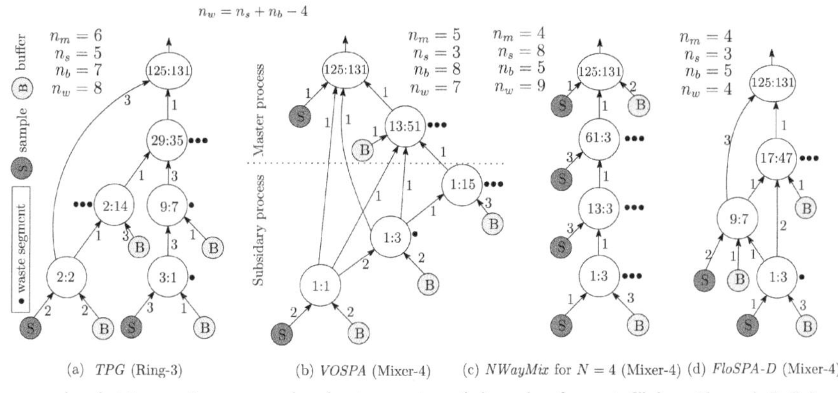

Figure 6.2 Generation of four units of target ratio (125:131) using sample preparation algorithms: (a) *TPG* [LSH15], (b) *VOSPA* [HLH15], (c) *NWayMix*, and (d) the proposed algorithm *FloSPA-D*.

[1]The detailed description of mixing models supported by Ring-N can be found in Section 2.4.

that is obtained based on the (1:1) mixing model, which is often used in digital microfluidics, e.g., in *twoWayMix* [TUTA08] or *REMIA* [LCH15]. The *TPG* algorithm applies tree-pruning and grafting on the initial dilution graph to obtain the dilution graph for a special type of unequally segmented rotary mixer (Ring-N [LSH15]). In each mixing step, *TPG* considers only a few specific volumetric ratios supported by $(k : \ell)$ mixing model ($k + \ell = 2^m$, $k, \ell, m \in \mathbb{N}$) supported by Ring-$N$. Therefore, *TPG* is unable to exploit all available mixing models for Ring-N. For example, in Ring-3, the possible mixing models are (1:1), (1:3), and (1:1:2); however, *TPG* uses only first two of them.

The volume-oriented sample preparation algorithm (*VOSPA* [HLH15]) uses a uniformly segmented rotary mixer (Mixer-N) and exploits the power of utilizing multiple intermediate *CF*s. *VOSPA* consists of two phases: master process and subsidiary process. The subsidiary process fills a concentration bank that can be used by the master process, which greedily accumulates various *CF*s from the bank for each segment of the mixer. Note that only one segment of Mixer-N is reused between two calls of the master process, and the remaining $(N - 1)$ segments are left unused. Moreover, the subsidiary process can use only two *CF*s in each mixing step for filling the bank that is subsequently used by the master process.

Note that both *TPG* and *VOSPA* produce a mixture of two reagents: sample and buffer. Algorithmic mixture preparation for more than two reagents using CFMB still remains unexplored. Proposed method also provides a solution to this general problem. Table 6.2 presents the scope of the proposed work against the background of prior art.

6.3 Motivation and Contribution

One of the major objectives in sample preparation is to minimize the assay time, which is proportional to the number of mixing operations [CUQ01]. Thies *et al.* [TUTA08] proposed a sample preparation algorithm (*MinMix*) for generating a target ratio based on the (1:1) mixing model. A special case of *MinMix* (known as *twoWayMix* [TUTA08] algorithm) can be used to perform dilution with at most d mixing steps, where d is the number of bits used to represent the numerator of the approximated target *CF*. In the case of Mixer-N, one can generalize *twoWayMix* for CFMBs so that it produces a given target *CF* based on multiple mixing models that are supported by Mixer-N; let us denote this procedure as *NWayMix*. The entire range of concentrations generated with Mixer-N can thus be represented as

Table 6.2 Scope of sample preparation algorithms on CFMB

Attribute	*TPG* [LSH15]	*VOSPA* [HLH15]	Proposed
Purpose	Reagent minimization.	Reagent minimization.	Reagent minimization using least-depth mixing graph.
Solution strategy	Iteratively modifies the dilution graph created by existing DMFB dilution algorithms.[1]	Greedy-heuristic based.	SMT-based optimal modeling.
Mixing models	Unable to exploit all mixing-ratios of Ring-*N*.	Only master process may use all mixing-ratios of Mixer-*N*.	Can utilize any mixing-ratio produced by Mixer-*N*, as needed.
# Mixing steps in dilution	Minimality not ensured.	Minimality not ensured.	Minimality guaranteed.
Mixture preparation	Not addressed.	Not addressed.	Reagent-saving mixture preparation.

[1] based on the (1:1) mixing model only.

$\{x : y \,|\, x + y = N^d\}$, where d digits in base-N are used to approximate a *CF* as $\frac{x}{N^d}$. Algorithm 1 describes the outline of *NWayMix*.

Example 6.3.1 Consider a mixing-ratio $\{\text{sample} : \text{buffer} = 125 : 131\}$, or equivalently, sample-*CF* in the target is $\frac{125}{256}$. Assume the availability of Mixer-4 ($N = 4$). Since the sum of ratio-components $125 + 131 = 256 = 4^4$, error-tolerance $\epsilon = 0$ can be achieved when d is chosen as 4. In Figure 6.2(c), the bottom most node indicates that one segment of the rotary Mixer-4 is filled with pure sample, whereas three remaining segments are filled with neutral buffer. Hence, the resultant *CF* of the sample in the mixture obtained after mixing is $\frac{1+3\times 0}{4} = \frac{1}{4}$. The *CF* of the buffer will be $1 - \frac{1}{4} = \frac{3}{4}$. Hence, for this intermediate fluid, $\{\text{sample} : \text{buffer} = 1 : 3\}$.

At the second mixing step (second node from bottom), three segments are filled with pure sample, and the remaining one segment is filled with the intermediate fluid obtained at the previous step. Hence, sample-*CF* will be $\frac{3+\frac{1}{4}}{4} = \frac{4\times 3+1}{4^2}$. Similarly, the sample-*CF* produced at the third mixing node is $\frac{4^2\times 3+4\times 3+1}{4^3}$. Finally, at the target node, it will become

Algorithm 1: *NWayMix*($\{x : y\}, N, d$)

Input: $\{$*sample* : *buffer* $= x : y\}$: Target ratio, N: Mixer-N
Output: Mixing steps
/* Represent x as d digits in base-N and $1 \leq x \leq N^d - 1$ */
$x = x_{d-1}x_{d-2}\ldots x_1x_0$, where $x_i \in \{0, 1, \ldots, N-1\}, 0 \leq i \leq d-1$;
/* Discard 0's, if any, from right to retain d' digits, $d' \leq d - 1$ */
$j = 0$;
while $x_j = 0$ **do**
    $j = j + 1$;
Fill x_j segments with *sample* and the remaining $(N - x_j)$ segments with *buffer*;
/* $Mix()$ denotes the mixing operation after filling all segments of Mixer-N */
$newFluid = Mix()$;
for $(j = j + 1; j \leq d - 1; j = j + 1)$ **do**
    Fill x_j segments with *sample* and $(N - x_j - 1)$ segments with *buffer*;
    Fill the last segment with $newFluid$;
    $newFluid = Mix()$;

$\frac{4^3 \times 1 + 4^2 \times 3 + 4^1 \times 3 + 4^0 \times 1}{4^4} = \frac{125}{4^4}$. Henceforth, for the sake of convenience, we will call sample-*CF* as target *CF*.

Algorithm 1 essentially expands 125 in base-4, i.e., $125 = (1331)_4 = 4^3 \times 1 + 4^2 \times 3 + 4^1 \times 3 + 4^0 \times 1$. Hence, conversely, the target *CF* can be achieved by filling the segments with pure sample in the same sequence as depicted in Figure 6.2(c). ■

The above argument can be generalized to justify the correctness of Algorithm 1. Given the target *CF*, N, and ϵ, the value of d is first derived. Next, the target *CF* is expressed as a fraction with at most d-digits in base-N. Since in each mixing step, only one segment of previously produced fluid is retained, the contribution of the pure sample added to a particular mixing step is reduced by a factor of N with each mixing step it undergoes while reaching the target. The process terminates after at most d mixing steps producing a *CF* value that exactly corresponds to the desired target ratio.

Note that $\frac{x}{N^d} (x, N, d \in \mathbb{N})$ may be reducible, i.e., there may be a common factor of N between the numerator and denominator. To express the ratio in irreducible form, one can simplify it as follows. Let $\frac{x}{N^d} = \frac{x'}{N^{d'}}$ $(x', d' \in \mathbb{N}, d' \leq d)$, where x' is not divisible by N ($x' \% N \neq 0$), i.e., the numerator and denominator are reduced until there does not exist any common factor of N between them. Note that, after reduction d'-bit representation of x' will have no trailing zeros in its base-N representation. Theorem 6.3.1, stated below, guarantees the minimality of the number of mixing steps required by *NWayMix*.

Theorem 6.3.1 Algorithm *NWayMix* generates a target ratio $\{\text{sample} : \text{buffer} = x : y\}$, where $x + y = N^d$, with minimum number of mixing operations d' ($d' \leq d$), when an N-segment rotary mixer (Mixer-N) is used.

Proof. By construction, it follows that the mixing graph is a tree, where the total number of internal nodes (representing mixing operations) is equal to the depth of the tree. Thus, the mixing tree resembles a chain (i.e., a skewed tree). Note that the target $CF = \frac{x}{N^d} = \frac{x'}{N^{d'}}$, where x' and $N^{d'}$ do not have any common factor of N. Hence, the depth of the tree is exactly d'. □

Although *NWayMix* generates a target CF using the minimum number of mixing operations, fluids residing in $(N - 1)$ segments are wasted in each mixing step when Mixer-N is used, as in the master process of *VOSPA* [HLH15]. Figure 6.2 shows the performance parameters of different sample preparation techniques for the target ratio $\{125 : 131\}$. It is observed that *NWayMix* produces the target CF using the fewest mixing operations but it consumes more units of reagents. On the other hand, *TPG* and *VOSPA* attempt to minimize reagent usage but may need more mixing steps, i.e., they may increase sample preparation time. A natural question in the context is: Can we reach a target CF with minimum mixing operations and then minimize reagent usage as a secondary objective? Figure 6.2(d) shows the dilution graph generated by the proposed algorithm *FloSPA-D*, which not only minimizes the number of mixing operations but also reduces reagent usage and waste production by cleverly sharing intermediate *CF*s. The following section briefly describes the overview of the solution procedure.

6.4 Overview of the Proposed Method

Proposed method models the dilution problem based on decision procedures over linear arithmetic and solves it by utilizing the deductive power of solvers for SMT [NOT06]. The SMT over linear arithmetic (SMT($\mathcal{LA}$)) problem is defined as follows.

Definition 6.4.1 Let Ψ be a first-order propositional formula, where the atoms are linear equations. The SMT($\mathcal{LA}$) problem is to determine an assignment to the variables of Ψ, if there exists a satisfiable assignment, otherwise to prove that no such assignment exists [KS08]. If a satisfiable assignment exists, we call the formula Ψ "satisfiable", otherwise Ψ is "unsatisfiable". ■

Example 6.4.1 Let $\Psi = (2x+3y \leq 5) \wedge (3x+5y \geq 6) \wedge (x \geq 0) \vee (y \leq 0)$. Here, Ψ is *satisfiable*, where $\{x = 2, y = 0\}$ is a satisfiable assignment. ■

Solving engines based on SMT algorithms have now reached enough sophistication, and hence they are being used for handling many complex optimization problems efficiently [KWHD14, KWD14, dMB11]. Although SMT solvers are used to solve a decision problem, an optimization problem can be formulated as a sequence of decision problems. In this context, the problem of diluting a sample is modeled using Mixer-N as a set of linear constraints over integers. Next, the SMT($\mathcal{LA}$) solver [dMB08] is invoked to check whether there exists a satisfiable assignment of underlying variables that can generate the target *CF* using only one unit of sample. If the answer is "yes", the desired solution is found. Otherwise, one needs to keep on increasing the number of sample-units and re-checking for satisfiability until success. Figure 6.3 shows the overview of the proposed dilution process. In the following sections, the optimization problem is formulated in detail.

6.5 Dilution

Proposed methodology exploits the fact that *NWayMix* generates a target concentration with the minimum number of mixing steps when Mixer-*N* is used. Starting with the mixing tree generated by *NWayMix*, an SAT-based technique is used to transform the tree into a mixing graph that corresponds to least reagent usage. Henceforth, the proposed method of Flow-based Sample Preparation Algorithm for Dilution is denoted as *FloSPA-D*. The dilution problem on a *CFMB* is now formally stated as follows:

- **Inputs:**
  - A supply of sample (*CF* $= 1$) and buffer (*CF* $= 0$);
  - A target concentration factor $C_t \in (0, 1)$;
  - An N-segment rotary mixer (Mixer-N);
  - A user-defined error-tolerance limit $\epsilon, 0 \leq \epsilon < 1$;
- **Output:**
  - Mixing graph for generating C_t within error limit ϵ starting with sample and buffer as input nodes;

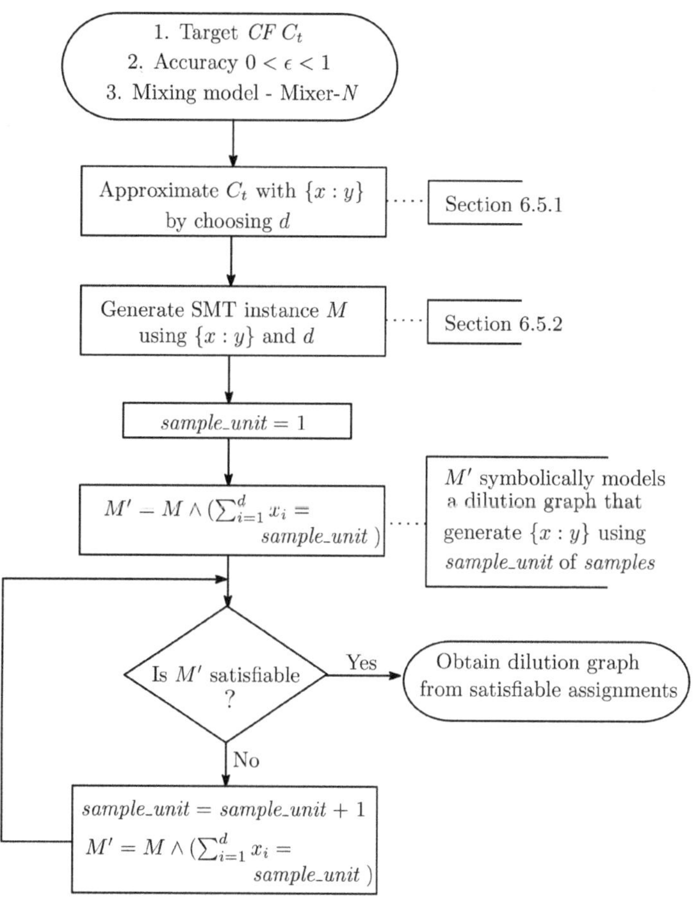

Figure 6.3 Proposed dilution (*FloSPA-D*) flow.

– **Objective:**

- Minimize the number of mixing operations as primary objective, and then minimize reagent usage as secondary objective.

6.5.1 Approximation of the Target Concentration Factor

Given a target CF $C_t \in (0, 1)$, error-tolerance ϵ, $0 \leq \epsilon < 1$ and the underlying mixing model supported by Mixer-N, the target CF is approximated as $C_t' = \frac{x}{N^d}$ (stated earlier in Section 6.1). Note that by Theorem 6.3.1, d also determines the depth of the skewed dilution tree, where the target node lies at depth 1.

Example 6.5.1 Let the desired *CF* $C_t = 0.39$, and a Mixer-4 is given, i.e., $N = 4$. Assume $\epsilon = 0.007$. Consider an approximation of C_t as $\{\text{sample} : \text{buffer} = 25 : 39\}$. Note that $|\frac{25}{4^3} - 0.39| = 0.0006 < \epsilon(= 0.007)$, for $d = 3$, which is the smallest value that satisfies the given error tolerance. Hence, the depth of the skewed dilution tree is 3. ■

6.5.2 Modeling of Dilution

The procedure *NWayMix* gives the minimum number of mixing steps for a given target ratio when Mixer-N is used. The resultant mixing tree is a chain (skewed) and each intermediate mixing step produces $(N - 1)$ waste segments. In the modeling, the reuse of intermediate fluids is maximized. The underlying search problem is modeled as a set of linear equations with logical connectives. For the sake of simplicity, the mixing graph is initially modeled as non-linear constraints over integers, and then it is systematically transformed into a set of linear constraints.

Modeling Formalism

Figure 6.4(a) shows the structure of the dilution graph for target ratio $\{X_1 : Y_1\}$, where $X_1 + Y_1 = N^d$. The dilution graph generated by *NWayMix* takes d mixing steps for generating $\{X_1 : Y_1\}$ using Mixer-N, i.e., the depth of the dilution graph is d. In order to identify how intermediate fluids can be reused, one need to transform the mixing tree generated by *NWayMix* by adding additional edges that represent such reuse of *CF*s. These edges represent possible sharing of intermediate *CF*s between non-adjacent mixing operations. In Figure 6.4(a), highlighted edges show the possibility of using intermediate *CF*s between non-consecutive mixing operations. Note that the knowledge of d is sufficient to generate the mixing graph corresponding to *FloSPA-D* while adding new edges on the skewed mixing tree. The value of d can be determined based on the error tolerance (ϵ) and N. Hence, *NWayMix* need not be run explicitly in order to create a *FloSPA-D* instance. Before going into the details, properties of the dilution graph are presented with reference to Figure 6.5.

Lemma 6.5.1 If k segments of a Mixer-N are filled with sample and are used in the mixing operation at depth i, where $1 \leq i \leq d$, they contribute the amount kN^{d-i} to the sample content X_i at this depth. The same is true for buffer also.

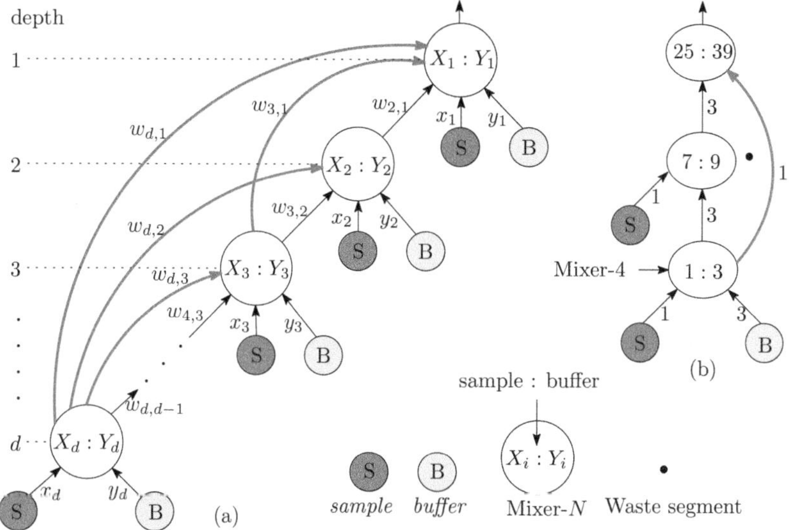

Figure 6.4 (a) Skeleton of a dilution graph for a target ratio $\{X_1 : Y_1\}$, where $X_1 + Y_1 = N^d$, using mixing models supported by Mixer-N and (b) the optimal solution for target ratio $\{25 : 39\}$ using Mixer-4.

Proof. By Theorem 6.3.1, the dilution graph is a tree of depth d. Note that, $X_i + Y_i = N^{d-i+1}$ for $1 \leq i \leq d$. Let k segments of the mixer at depth i be filled with the sample. Without loss of generality, one can assume that one segment of the Mixer-N at depth i is filled with the intermediate fluid obtained at preceding mixing operation at depth $i+1$, and the remaining $(N-k-1)$ segments are filled with buffer, or with intermediate *CF*s from previous operations at depth $i+2, i+3, \ldots, d$. Figure 6.5(a) shows the skeleton of the dilution graph in this setting. Hence, the portion (X_i) of the sample in the resultant mixing operation executed at depth i can be calculated as shown below, where $(\ldots)$ denotes the contributions of sample to X_i received from other incoming edges (dotted lines in Figure 6.5(a)):

$$X_i = \frac{\frac{X_{i+1}}{N^{d-i}} + k + (\cdots)}{N}$$
$$= \frac{(\cdots) + kN^{d-i} + (\cdots)}{N^{d-i+1}}$$

Hence, an amount of kN^{d-i} is contributed to X_i. □

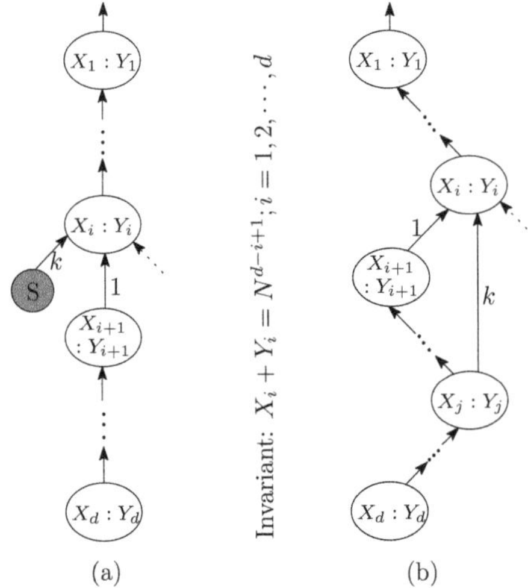

Figure 6.5 Skeleton of a dilution graph for proving the statements in (a) Lemma 6.5.1 and (b) Lemma 6.5.2.

Example 6.5.2 Consider a mixing node at depth 2 in Figure 6.2(c) with concentration ratio {sample : buffer = 61 : 3}. Note that target ratio is at depth 1. Three segments are filled with sample i.e., $k = 3$, and the remaining segment is filled with fluids from mixing node at depth 3. Hence, the portion of sample in the mixing node at depth 2 is $\frac{3+\frac{13}{4^2}}{4} = \frac{3\times4^2+13}{4^2\times4} = \frac{61}{4^3}$. Note that three units of sample and one unit of intermediate fluid reused from the mixing node at depth 3, contribute $3 \times 16 = 48$ and 13, respectively, to produce 61 at depth 2. ■

Lemma 6.5.2 If the fluid in k segments of Mixer-*N* at depth j is shared at depth i, where $1 \leq i < j \leq d$, then it contributes an amount $N^{j-i-1}X_j$ and $N^{j-i-1}Y_j$ to X_i and Y_i, respectively.

Proof. By Theorem 6.3.1, the dilution graph is a tree of depth d. Note that $X_i+Y_i = N^{d-i+1}$ for $1 \leq i \leq d$. Let k segments of the rotary mixer at depth i be filled with the intermediate fluid produced after the mixing operation at depth j, where $j > i$. Without loss of generality, one can assume that a segment of the rotary mixer at depth i is filled with the intermediate fluid at depth $i+1$, and the remaining $(N-k-1)$ segments are filled with sample, buffer, or with intermediate *CF*s produced at depth $i+2, i+3, \ldots, j-1, j+1, \ldots, d$. Figure 6.5(b) shows the skeleton of the dilution graph in this setting.

The portion of the sample in the resultant mixing operation at depth i, i.e., X_i can be calculated as follows:

$$X_i = \frac{\frac{X_{i+1}}{N^{d-i}} + k\frac{X_j}{N^{d-j+1}} + (\cdots)}{N}$$
$$= \frac{X_{i+1} + kN^{(d-i)-(d-j-1)}X_j + (\cdots)}{N^{d-i+1}}$$
$$= \frac{(\cdots) + kN^{j-i-1}X_j + (\cdots)}{N^{d-i+1}}$$

Thus, the mixing operation at depth j contributes the amount $kN^{j-i-1}X_j$ towards X_i at depth i when k segments are filled with intermediate fluids produced at depth j. □

Example 6.5.3 Let us consider a mixing node at depth 2 in Figure 6.2(d), where $\{$*sample* : *buffer* = $17 : 47\}$. Each of the two segments is filled with *buffer* and intermediate fluids obtained from the mixing operation at depth 3. The remaining two segments are filled with fluids from the mixing node at depth 4. Hence, the portion of *sample* in the mixing node at depth 2 is $\frac{\frac{9}{4^2}+2\times\frac{1}{4}}{4} = \frac{\frac{9+2\times 4}{4^2}}{4} = \frac{17}{4^3}$, i.e., when each unit of fluid from the mixing node at depth 4 is used in the mixing node at depth 2, it contributes the amount $4^{(4-2-1)} = 4$ to the *sample* content in the resultant concentration. ■

SMT-based Modeling

In the proposed modeling, several variables are defined. Figure 6.4(a) shows different types of variables as edge labels. The non-negative integer variables are defined as follows.

Node variables: For each mixing node at depth i in the dilution graph, two variables X_i and Y_i ($1 \leq i \leq d$) are defined that denote the ratio between sample and buffer in the intermediate fluid produced by the mixing operation at depth i.

Reagent variables: The input reagents (sample and buffer) can be used in any mixing node at depth i, where $1 \leq i \leq d$. For denoting the number of segments filled with sample and buffer in a Mixer-N at depth i, two variables x_i and y_i are associated, respectively.

Segment-sharing variables: The integer variable $w_{i,j}$ represents the number of segments that are used to fill Mixer-N at depth j from the intermediate fluid produced by Mixer-N at depth i, where $1 \leq j < i \leq d$.

Example 6.5.4 For the target ratio $\{25 : 39\}$, the depth (d) of the dilution graph is three. Therefore, the exploration of first three depths (1–3) of the dilution graph (shown in Figure 6.4(a)) suffices for our search. For $1 \leq i \leq 3$, the *node variables* are X_i, Y_i, and the *reagent variables* are x_i, y_i. Additionally, $w_{3,2}, w_{3,1}$, and $w_{2,1}$ denote *segment-sharing variables.* ■

The correctness of mixing ratios generated at each node is guaranteed by the following equations that model the consistency of a mixing node at depth j for $\forall j, 1 \leq j \leq d$.

Note that the ratio of sample and buffer at the mixing node at depth j is $\{X_j : Y_j\}$. From Figure 6.4(a), it can be seen that the mixing node at depth j can fill the segments of Mixer-N with sample, buffer, or with any unused fluid segments produced at depth $i > j$. From Lemma 6.5.1, it follows that sample or buffer contributes N^{d-j} to Mixer-N at depth j. Moreover, Lemma 6.5.2 implies that if $1 \leq j < i \leq d$, then fluids produced at depth i will contribute $N^{i-j-1}X_i$ and $N^{i-j-1}Y_i$ to X_j and Y_j, respectively, at node j. Therefore, the following equations should be satisfied for the desired ratio $X_j : Y_j$ at depth j.

$$N^{d-j}x_j + \sum_{i=j+1}^{d} N^{i-j-1}X_i w_{i,j} = X_j \tag{6.1}$$

$$N^{d-j}y_j + \sum_{i=j+1}^{d} N^{i-j-1}Y_i w_{i,j} = Y_j \tag{6.2}$$

Note that $X_j + Y_j = N^{d-j+1}$ for $j = 1, 2, \ldots, d$, where X_1 (Y_1) denotes the portion of sample (buffer) in the target.

Example 6.5.5 Consider the target ratio $\{25 : 39\}$. In the corresponding mixing graph, at depth 1, the required constraints are $16x_1 + 4X_3w_{3,1} + X_2w_{2,1} = 25, 16y_1 + 4Y_3w_{3,1} + Y_2w_{2,1} = 39$. Similarly, at depth 2, we have $4x_2 + X_2w_{3,2} = X_2, 4y_2 + Y_2w_{3,2} = Y_2$, and at depth 3, $X_3 = x_3, Y_3 = y_3$. ■

Moreover, we need to ensure that all N segments for Mixer-N be filled with intermediate *CF*s or input reagents; also Mixer-N can feed at most N segments of other mixers. Therefore, the consistency conditions at depth j are given by:

$$x_j + y_j + \sum_{i=j+1}^{d} w_{i,j} = N, \quad \sum_{i=1}^{j-1} w_{j,i} \leq N \tag{6.3}$$

Finally, all weights must satisfy $0 \leq w_{i,j} \leq N-1$, for $1 \leq j < i \leq d$. Similarly, $0 \leq x_i, y_i \leq N-1$ for $1 \leq i \leq d$.

Example 6.5.6 The consistency conditions for depth 1 are $w_{3,1}+w_{2,1}+x_1+y_1 = 4$; edge weights must satisfy $0 \leq w_{3,1}, w_{2,1}, x_1, y_1 \leq 3$. For depth 2, $w_{3,2}+x_2+y_2 = 4, w_{2,1} \leq 4$ and $0 \leq w_{3,2}, x_2, y_2 \leq 3$. Finally, for depth 3, the constraints are $x_3+y_3 = 4, w_{3,1}+w_{3,2} \leq 4$, and $0 \leq x_3, y_3 \leq 3$. ■

Elimination of Non-linearity

Note that the constraints in Eqn. (1) and Eqn. (2) are nonlinear in nature. Unfortunately, the polynomial constraint solving problem over the integers is undecidable [NOT06]. For the purpose of transforming nonlinear equations into SMT($\mathcal{LA}$), the inherent properties of the underlying mixing graph is exploited to construct a modeling proportional to its size. Note that $w_{i,j}$ is non-negative and bounded by the segment size of Mixer-N (in practice , $N \leq 8$), i.e., $0 \leq w_{i,j} \leq N-1$. Hence, the non-linearity in Eqn. (6.1) and Eqn. (6.2) can be dealt as follows. For each $X_i w_{i,j}$ in Eqn. (6.1), we replace it with non-negative integral variable $X'_{i,j}$ and add the following constraints (Eqn. (6.2) can also be handled similarly):

$$(w_{i,j} = k) \implies (X'_{i,j} = kX_i), \text{ for } k = 0, 1, \ldots, N-1 \tag{6.4}$$

Example 6.5.7 Consider the non-linear equation $16x_1 + 4X_3w_{3,1} + X_2w_{2,1} = 25$ in Example 6.5.5. The given equation is transformed into equivalent linear equations as follows: $16x_1 + 4X'_{3,1} + X'_{2,1} = 25$ and $((w_{3,1} = k) \implies (X'_{3,1} = kX_3)) \wedge ((w_{2,1} = k) \implies (X'_{2,1} = kX_2))$, $k = 0, 1, 2, 3$. ■

6.5.3 Dilution Algorithm

For a given target *CF* C_t and an error tolerance ϵ, we get an approximated target ratio $\{x : y\}$ along with the depth of dilution graph d (line 1 of Algorithm 2), which is same as the number of mixing operations in *NWayMix*. Hence, the optimality in the number of mixing steps is guaranteed (by Theorem 6.3.1). However, to minimize reagent consumption, we have adopted an iterative solution approach as outlined in Algorithm 2 (lines 2–9). Line 2 generates the SMT instance (as discussed in Section 6.5.2) for the desired dilution graph of depth d and let it be M. Next, we check the existence of a dilution graph that takes one unit of sample. Thus, we add

Algorithm 2: *FloSPA-D*(C_t, ϵ, N)

Input: C_t: target CF, ϵ: accuracy, N: Mixer-N
Output: Dilution Graph

1 Approximate C_t by choosing the smallest $d \in \mathbb{N}$ such that $|\frac{x}{x+y} - C_t| < \epsilon$; Target ratio is $\{x : y\}$, where $x + y = N^d$ and $1 \leq x \leq N^d - 1$;
/* Detailed modeling is discussed in Section 6.5.2 */
2 M = SMT instance generated from a dilution graph of depth d and$\{x : y\}$;
/* one sample_unit is equal to the volume of 'reagent' in one segment of rotary mixer */
3 *sample_unit* $= 1$;
4 $M' = M \wedge (\sum_{i=1}^{d} x_i =$ *sample_unit*$)$;
5 $checkSAT(M')$;
6 **while** M' *is* unsatisfiable **do**
7   *sample_unit* = *sample_unit* + 1;
8     $M' = M \wedge (\sum_{i=1}^{d} x_i =$ *sample_unit*$)$;
9     $checkSAT(M')$;
10 Obtain dilution graph from "satisfiable" assignments of M';
11 **return** dilution graph;

$\sum_{i=1}^{d} x_i = 1$ to M obtain M' (line 3–4), and check the satisfiability of M' (line 5). If M' is "satisfiable", we obtain a solution. Otherwise, we need to check for the existence of a dilution graph that needs one additional unit of sample and re-check for satisfiability (line 6–9). This process of adding more sample units is continued until M' is observed to be "satisfiable". Hence, it guarantees "optimal sample consumption" under the constraint of having minimum number of mixing steps. Figure 6.4(b) shows the final solution generated by *FloSPA-D* for the target ratio $\{25 : 39\}$. The following theorem formalizes the optimality of the flow-based dilution algorithm *FloSPA-D*.

Theorem 6.5.3 *FloSPA-D* generates a target *CF* C_t, $0 < C_t < 1$ within the given error limit $\epsilon, 0 \leq \epsilon < 1$ that minimizes reagent usage under the constraint of using minimum number of mixing steps deployed by Mixer-*N*.

Proof. Theorem 6.3.1 guarantees the optimality of the number of mixing operations needed in the dilution tree returned by *NWayMix*. For maximal reuse of intermediate *CFs* produced by *NWayMix*, *FloSPA-D* encodes the dilution tree with a set of linear equations. The correctness of the encoding scheme is guaranteed by Lemma 6.5.1 and Lemma 6.5.2. Note that decision procedures of the SMT-over-linear-arithmetic is sound and complete [KS08]. We start with one unit of sample and invoke the SMT-solver. If the answer is "yes", we obtain a feasible solution. Otherwise, the number of sample-units is increased in steps, and checking for satisfiability is continued until we are successful. Hence, the consumption of minimum amount of reagent is ensured in the final solution. □

6.6 Mixture Preparation

In this section, we discuss how an SMT-based approach can be extended to implement the more general problem of mixture preparation. We first modify the *MinMix* algorithm (valid only for (1:1) mixing model [TUTA08]) so that it can generate a mixing tree under the generalized mixing models supported by Mixer-N; let us denote it as *genMixing*. Our objective is to produce a target ratio $\mathcal{M} = \{\langle R_1, c_1\rangle, \langle R_2, c_2\rangle, \ldots, \langle R_k, c_k\rangle\}$ following a sequence of mixing operations supported by Mixer-N.

6.6.1 Approximation of the Target Mixture-Ratio

Given a target mixture $\mathcal{M}$ of k reagents with an error-tolerance $\epsilon, 0 \leq \epsilon < 1$, and a Mixer-$N$, we approximate the target mixture $\mathcal{M}$ with $\mathcal{R} = \{x_1 : x_2 : \ldots : x_k\}$ by choosing the smallest $d \in \mathbb{N}$ such that $\max_i\{|c_i - \frac{x_i}{N^d}|\} < \epsilon$ and $\sum_{i=1}^{k} x_i = N^d$, where d is the depth of the mixing tree representing the sequence of mixing operations that lead to $\mathcal{M}$.

Example 6.6.1 Consider a mixture $\mathcal{M} = \{\langle R_1, 0.30\rangle, \langle R_2, 0.23\rangle, \langle R_3, 0.24\rangle, \langle R_4, 0.23\rangle\}$ and assume that Mixer-4 ($N = 4$) is used. Assume that the error limit is $\epsilon = 0.007$. $\mathcal{M}$ can be approximated as $\{R_1 : R_2 : R_3 : R_4 = 19 : 15 : 15 : 15\}$, since $\max\{|0.30 - \frac{19}{4^3}|, |0.23 - \frac{15}{4^3}|, |0.24 - \frac{15}{4^3}|\} = 0.005 < \epsilon$, and $19 + 15 + 15 + 15 = 4^3$, i.e., a choice of $d = 3$ suffices here. Hence, the depth (d) of the mixing tree is 3. ■

6.6.2 Generalized Mixing Algorithm

In this subsection, we present the algorithm (*genMixing*) that will produce a target ratio using Mixer-N.

Example 6.6.2 Figure 6.6 shows the resultant mixing tree for the target ratio considered in Example 6.6.1. Initially, *genMixing* creates $d+1$ empty bins/stacks $Bin_0, Bin_1, \ldots, Bin_d$. Next, it represents each input reagent R_i with d digits using the base-N number system;

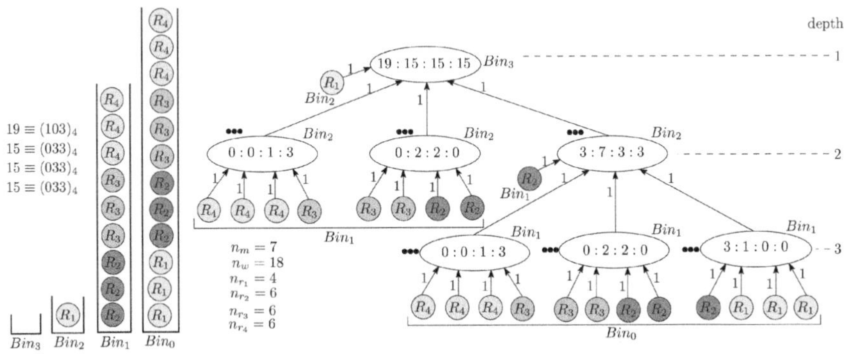

Figure 6.6 Mixing tree for the approximated target ratio $\{R_1 : R_2 : R_3 : R_4 = 19 : 15 : 15 : 15\}$ produced by *genMixing*.

let $R_i = (a^i_{d-1}a^i_{d-2}\ldots a^i_1 a^i_0)_N$. In the next step, *genMixing* scans each digit from right to left and populate Bin_j with a^i_j units of R_i for $0 \leq j \leq d-1$. Note that $R_1 = 19$ is represented as 103 in base-4. Hence, Bin_2 and Bin_0 are populated with 1 and 3 units (1 unit = 1 segment of Mixer-4) of reagent R_1, respectively. Finally, *genMixing* invokes another function *genMixingHelper* (Algorithm 4) recursively, to generate the desired mixing tree.

The function *genMixingHelper* invokes the use of bins starting from Bin_d recursively. When a bin is empty, an internal node is created into the mixing

---

Algorithm 3: *genMixing*$(\mathcal{M}, \epsilon, N)$

Input: $\mathcal{M}$: target mixture with k input reagents, ϵ: accuracy, N: Mixer-N
Output: Mixing tree

Approximate $\mathcal{M}$ with $\mathcal{R} = x_1 : x_2 : \ldots : x_k$ by choosing the smallest d so that $\max_{i\in\{1,2,\ldots,k\}}\{|c_i - \frac{x_i}{N^d}|\} < \epsilon$ and $\sum_{i=1}^{k} x_i = N^d$;
Create an array of $d+1$ empty bins $\mathcal{B} = [Bin_0, Bin_1, \ldots, Bin_d]$;
for $(i = 1; i \leq k; i = i + 1)$ **do**
    /* Represent R_i as d-digit base-N number */
    Let $R_i = (a^i_{d-1}a^i_{d-2}\ldots a^i_1 a^i_0)_N$;
    for $(j = 0; j < d; j = j + 1)$ **do**
        /* push a^i_j copies of R_i at Bin_j */
        for $(l = 1; l \leq a^i_j; l = l + 1)$ **do**
            $\mathcal{B}[j].push(R_i)$;

return genMixingHelper$(\mathcal{B}, d)$;

Algorithm 4: *genMixingHelper*$(\mathcal{B}, d)$

Input: $\mathcal{B}$: An array of $d+1$ bins, d: depth of mixing tree
Output: Mixing tree

if $\mathcal{B}[d] = \phi$ **then**
    for $(i = 1; i \leq N; i = i + 1)$ **do**
        T_i =*genMixingHelper*$(\mathcal{B}, d-1)$;
    Create an internal node T with $\langle T_1, T_2, \ldots, T_N \rangle$ as its children;
    return T;
else
    $R = \mathcal{B}[d].pop()$;
    Create a leaf node $\mathcal{L}$ with a value of R;
    return $\mathcal{L}$;

tree. Also, N recursive procedures are invoked through Bin_{d-1} for filling N segments of the mixer corresponding to the internal node. Otherwise, it pops one reagent-unit from the non-empty bin and returns it as a leaf/reagent node in the mixing tree. Here, *genMixing* calls *genMixingHelper* with $d = 3$. As Bin_3 is empty, four calls are generated recursively for Bin_2. The first call of *genMixingHelper* for Bin_2 finds it non-empty. Hence, a leaf node with a reagent R_1 is created and attached to the internal node created for Bin_3 (root node in Figure 6.6). For all three remaining calls, Bin_2 is empty. The procedure *genMixingHelper* returns the desired mixing tree as shown in Figure 6.6, where each node is labeled with Bin_i. ■

Note that *genMixing* produces a mixing tree of depth d, given $\mathcal{M}$, ϵ, and N. The *CF* of each reagent R_i is expressed as $\frac{x_i}{N^d}$, where $x_i \in \mathbb{N}$. Consider the bottom-up realization of the mixing tree (Figure 6.6). In each level, the *CF* of the input reagents is reduced by a factor of N. Thus, in order to achieve N^d in the denominator of *CF*, at least d sequential mixing steps are needed. Algorithm 3 exactly produces a mixing tree of depth d, and hence, it is of minimum depth (height).

6.6.3 SMT-based Modeling of Reagent-saving Mixing

In *genMixing*, fluids residing in each of $(N-1)$ segments of the internal mixing nodes are wasted, and hence, the overall reagent consumption may become high. Hence, in contrast to *FloSPA-D*, maximizing the sharing of waste segments may not be sufficient for reducing reagent usage in this case.

We need to reduce the number of mixing operations by sharing intermediate nodes as far as possible. The proposed modeling that incorporates these issues is described as follows.

Symbolic Formulation

We create a skeleton tree by adding all missing reagent leaf nodes to all internal/mixing nodes of the mixing tree returned by *genMixing*. The skeleton tree for Figure 6.6 is shown in Figure 6.7(a). Next, we encode it symbolically using a set of non-negative integer variables and linear equations. The encoding is somewhat similar to the modeling of dilution; however, some additional variables are required for handing multiple reagents. For the ease of illustration, we describe the detailed modeling for the target ratio $\{R_1 : R_2 : R_3 : R_4 = 19 : 15 : 15 : 15\}$. In the proposed modeling, we define several non-negative integer variables as follows.

Node variables: For every internal/mixing node of the skeleton tree, k node variables $R_1^l(i), R_2^l(i), \ldots, R_k^l(i)$ are assigned for denoting the portion of input reagents $R_1, R_2, \ldots, R_k$, respectively, where l is the depth of that node and i is the index of the node at depth l. Note that at each depth, the index of a mixing node is assigned from left to right starting with 1. For example, the node variables corresponding to the rightmost node of depth 2 in Figure 6.7(a) are $R_1^2(3), R_2^2(3), R_3^2(3), R_4^2(3)$.

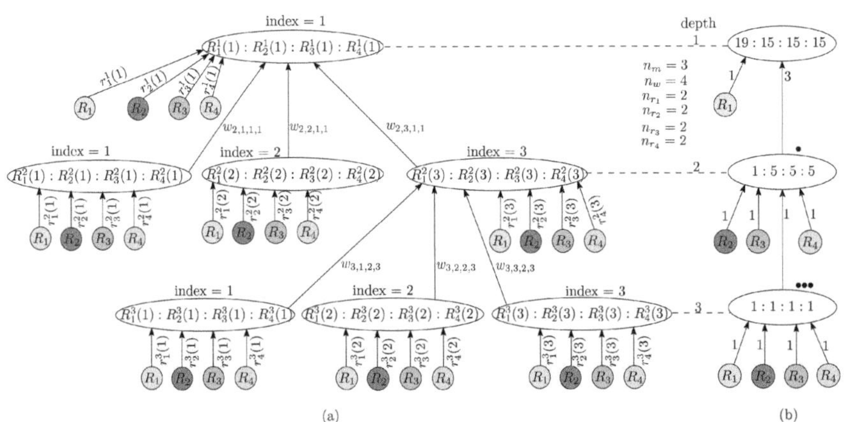

Figure 6.7 (a) Skeleton tree obtained from the mixing tree (Figure 6.6). (b) Reagent-saving mixing tree obtained by *FloSPA-M*.

Reagent variables: We define reagent variables $r_j^l(i)$, where $j = 1, 2, \ldots, k$, to denote the number of segments filled with reagent R_j, where l is the depth of the mixing node where reagent R_j is used, and i is the index of the mixing node. For example, the reagent variables corresponding to the rightmost node on depth 2 in Figure 6.7(a) are $r_1^2(3), r_2^2(3), r_3^2(3), r_4^2(3)$.

Segment-sharing variables: An edge (u, v) between two mixing nodes denotes the sharing of *CF* from node u to node v in the mixing tree. An edge variable w_{l_1,i_1,l_2,i_2} denotes the number of segments shared between a node with index i_1 at depth l_1, and a node with index i_2 at depth l_2. For example, edge weight $w_{3,3,2,3}$ in Figure 6.7(a) represents the number of segments shared from the node indexed as 3 at depth 3 (the rightmost node at depth 3) to the node with index 3 at depth 2 (the rightmost node at depth 2).

For simplicity, we assume that intermediate *CF*s are shared between two adjacent depths only. This restriction can be relaxed by suitably modifying the constraints for each mixing node. In order to ensure the correctness of the mixing ratio generated at each mixing node, let us consider a mixing node (l, i) (denoting the mixing node at depth l with index i) that has k reagent nodes as leaves and m internal mixing nodes (at depth $l + 1$) as its children. Figure 6.8 shows the portion of the subtree rooted at (l, i). Let the height of m subtrees of the mixing node (l, i) be $h_1, h_2, \ldots, h_m$. Hence, the height of the entire subtree shown in Figure 6.8 is $h = \max\{h_1, h_2, \ldots, h_m\} + 1$. The desired ratio $R_1^l(i) : R_2^l(i) : \ldots : R_k^l(i)$ can be calculated with the following non-linear equations, which can be transformed into a set of equivalent linear equations.

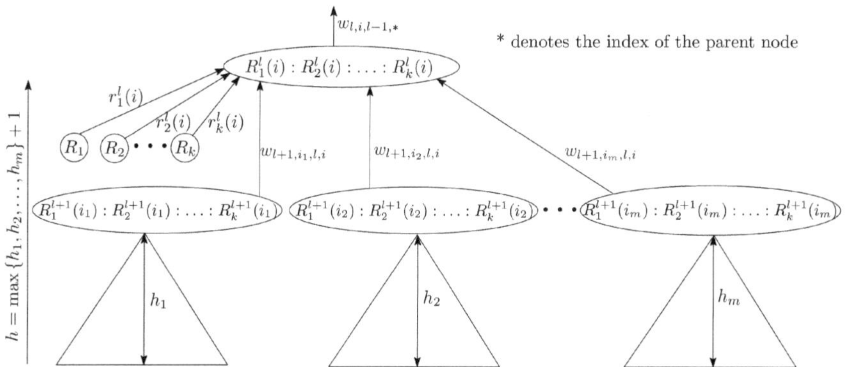

Figure 6.8 Structure of a subtree in the skeleton mixing tree.

$$
\begin{aligned}
N^{h-2}r_1^l(i) + \sum_{j\in\{i_1,i_2,\ldots,i_m\}} N^{h-h_j-1}w_{l+1,j,l,i}R_1^{l+1}(j) &= R_1^l(i)\\
N^{h-2}r_2^l(i) + \sum_{j\in\{i_1,i_2,\ldots,i_m\}} N^{h-h_j-1}w_{l+1,j,l,i}R_2^{l+1}(j) &= R_2^l(i)\\
\ddots \qquad\qquad\qquad\qquad &\quad \vdots\\
N^{h-2}r_k^l(i) + \sum_{j\in\{i_1,i_2,\ldots,i_m\}} N^{h-h_j-1}w_{l+1,j,l,i}R_k^{l+1}(j) &= R_k^l(i) \qquad (6.5)
\end{aligned}
$$

Moreover, we need to ensure that either all N input segments for a Mixer-N are filled with intermediate fluids/input reagents or the concerned mixing operations are not necessary for generating the desired target ratio. In the second case, the number of mixing operations is reduced. A Mixer-N can provide fluids to at most N segments of other mixers. The required consistency conditions are encoded as follows:

$$
\left(\sum_{j=1}^{k} r_j^l(i) + \sum_{j\in\{i_1,i_2,\ldots,i_m\}} w_{l+1,j,l,i} = N\right) \bigvee
$$
$$
\left(\sum_{j=1}^{k} r_j^l(i) + \sum_{j\in\{i_1,i_2,\ldots,i_m\}} w_{l+1,j,l,i} = 0\right),
$$
$$
0 \le w_{l,i,l-1,*} \le N-1 \qquad (6.6)
$$

All weights must satisfy $0 \le w_{l+1,j,l,i} \le N-1$, for $j \in \{i_1, i_2, \ldots, i_m\}$. Analogously, $0 \le r_j^l(i) \le N-1$ for $j = 1, 2, \ldots, k$.

The following example illustrates the consistency constraints for the root node of Figure 6.7(a).

Example 6.6.3 The root node of the tree (Figure 6.7(a)) has three incoming edges from other internal nodes. The depths of its subtrees are 2, 2, and 3 in the left-to-right order. Moreover, the depth of the complete tree, i.e., h, is 4. The consistency conditions for the root node in Figure 6.7(a) are given as follows.

$$
\begin{aligned}
4^2r_1^1(1) + 4w_{2,1,1,1}R_1^2(1) + 4w_{2,2,1,1}R_1^2(2) + w_{2,3,1,1}R_1^2(3) &= R_1^1(1)\\
4^2r_2^1(1) + 4w_{2,1,1,1}R_2^2(1) + 4w_{2,2,1,1}R_2^2(2) + w_{2,3,1,1}R_2^2(3) &= R_2^1(1)
\end{aligned}
$$

$$4^2 r_3^1(1) + 4w_{2,1,1,1}R_3^2(1) + 4w_{2,2,1,1}R_3^2(2) + w_{2,3,1,1}R_3^2(3) = R_3^1(1)$$
$$4^2 r_4^1(1) + 4w_{2,1,1,1}R_4^2(1) + 4w_{2,2,1,1}R_4^2(2) + w_{2,3,1,1}R_4^2(3) = R_4^1(1)$$
$$\left(r_1^1(1) + r_2^1(1) + r_3^1(1) + r_4^1(1) + w_{2,1,1,1} + w_{2,2,1,1} + w_{2,3,1,1} = 4\right) \bigvee$$
$$\left(r_1^1(1) + r_2^1(1) + r_3^1(1) + r_4^1(1) + w_{2,1,1,1} + w_{2,2,1,1} + w_{2,3,1,1} = 0\right)$$

Additionally, $0 \leq r_1^1(1), r_2^1(1), r_3^1(1), r_4^1(1) \leq 3$ and $0 \leq w_{2,1,1,1}, w_{2,2,1,1}, w_{2,3,1,1} \leq 3$. ■

6.6.4 Reagent-Saving Mixing Algorithm

In summary, for a given $(\mathcal{M}, \epsilon, N)$, a mixing tree is first generated using *genMixing*. In the next step, the skeleton tree is created. An SMT instance for the skeleton tree is then generated as discussed in Section 6.6.3. All non-linearities in the constraints are removed using certain transformations as described earlier. Let the set of constraints be M. Next, we check the existence of a mixing tree that takes N units of reagents. Thus, we add $\sum_{x \in \text{set of all reagent variables}} x = N$ to M and obtain M'. If M' is "satisfiable", a solution is found. Otherwise, we need to check for the existence of a mixing tree assuming the allowance for one additional unit of reagent and re-check for satisfiability. The selection of additional reagent is automatically decided by the SAT-solver based on the underlying constraints. This process is continued until M' is found to be "satisfiable". Algorithm 5 summarizes the steps needed in the Flow-based Sample Preparation Algorithm for Mixtures (*FloSPA-M*).

Example 6.6.4 The skeleton tree derived from *genMixing* and the mixing tree for *FloSPA-M* are shown in Figure 6.7(a) and Figure 6.7(b), respectively. Note that the number of mixing steps and waste segments are 7 and 18 for *genMixing* (Figure 6.6), whereas *FloSPA-M* produces the same mixture (Figure 6.7(b)) with only three mixing steps and four waste segments. Moreover, the total reagent consumption in *FloSPA-D* (8) is significantly less than that required in *genMixing* (22). Table 6.3 summarizes the performance of *genMixing* and *FloSPA-M*. ■

We further improve the *FloSPA-M* algorithm so that the intermediate ratios produced at different nodes of the mixing tree can be used not only by its parent but by all ancestors in the mixing tree, thereby enhancing the possibility of sharing intermediate fluids compared to *FloSPA-M*. We call this algorithm as "*FloSPA* for Enhanced Mixing" (*FloSPA-EM*). In

Algorithm 5: *FloSPA-M*$(\mathcal{M}, \epsilon, N)$

Input: $\mathcal{M}$: target mixture, ϵ: accuracy, N: Mixer-N
Output: Mixing Tree

1 $T = genMixing(\mathcal{M}, \epsilon, N)$;
2 T_1 = Skeleton tree of T after adding reagent nodes;
/* Detailed modeling and removal of non-linearity are discussed in Section 6.6.3 and Section 6.5.2, respectively */
3 M = SMT instance generated from T_1 after removing non-linearity;
/* At least N reagent units are required for filling N segments of a Mixer-N */
4 $total\_reagent\_unit = N$;
/* total_reagent_unit denotes the summation of all input reagents */
5 $M' = M \wedge (\sum_{x \in \text{set of all reagent variables}} x = reagent\_unit)$;
6 $checkSAT(M')$;
7 **while** M' *is "unsatisfiable"* **do**
8     $total\_reagent\_unit = total\_reagent\_unit + 1$;
9     $M' = M \wedge (\sum_{x \in \text{Set of all reagent variables}} x = total\_reagent\_unit)$;
10     $checkSAT(M')$;
11 Obtain mixing tree from "satisfiable" assignments of M';
12 **return** mixing tree;

Table 6.3 Performance comparison between *genMixing* and *FloSPA-M* for $\{19 : 15 : 15 : 15\}$

Parameter	*genMixing*	*FloSPA-M*	%-savings
n_m	7	3	57.14%
n_w	18	4	77.77%
n_{r_1}	4	2	50.00%
n_{r_2}	6	2	66.67%
n_{r_3}	6	2	66.67%
n_{r_4}	6	2	66.67%

FloSPA-EM, we add a few additional edges in the skeleton tree that represent possible sharing of intermediate fluids between a node and its ancestors in the skeleton tree while constructing the "enhanced skeleton tree". For implementing of *FloSPA-EM*, only line-2 of Algorithm 5 needs to be modified by

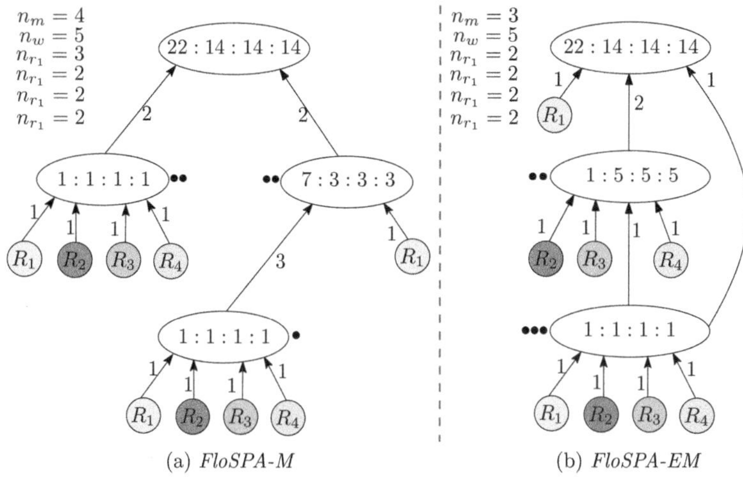

Figure 6.9 Reagent-saving mixing tree obtained by (a) *FloSPA-M* and (b) *FloSPA-EM*.

including "enhanced skeleton tree" in place of skeleton tree, keeping the rest of the codes unchanged.

Example 6.6.5 When we run *FloSPA-M* and *FloSPA-EM* for the target ratio $\{R_1 : R_2 : R_3 : R_4 = 22 : 14 : 14 : 14\}$, *FloSPA-EM* performs better compared to *FloSPA-M*. The corresponding mixing trees for *FloSPA-M* and *FloSPA-EM* are shown in Figure 6.9(a) and Figure 6.9(b), respectively. ■

6.7 Experimental Results

We have implemented, in Python, the proposed dilution algorithms *FloSPA-D* and *NWayMix* and compared their performance with *VOSPA* [HLH15] and *TPG* [LSH15]. We have also evaluated the proposed mixture preparation algorithms *genMixing, FloSPA-M*, and *FloSPA-EM*. The clauses are represented using the Python library Z3Py that can handle linear constraints with Boolean connectives for the SMT-solver Z3 [dMB08]. As an implementation platform, we have used 2 GHz Intel Core i5 computer with 8 GB memory running 64-bit Ubuntu 14.04 operating system.

6.7.1 Performance Evaluation for Dilution

We have compared the proposed dilution technique with two state-of-the-art sample preparation algorithms (*VOSPA* [HLH15], *TPG*[2] [LSH15]), and also with the baseline algorithm *NWayMix* using Mixer-N. We have conducted three sets of experiments. In the first set, we use a four-segment mixer (Mixer-4) to generate all CFs ranging from $\frac{1}{256}$ to $\frac{255}{256}$ (i.e., $d = 4, N = 4$) and report the average number of mixing steps ($\bar{n}_m$), the average number of sample ($\bar{n}_s$) and buffer ($\bar{n}_b$) units consumed, and the average number of wasted units ($\bar{n}_w$) while producing a target CF. These parameter are shown in Table 6.4 in which %-improvement of *FloSPA-D* compared to other methods are shown in parenthesis. The formula used for calculating savings is shown in the footnote of Table 6.4. In the remaining two sets of experiments, we increase the accuracy (ϵ) of the target CFs and use more general mixing model, i.e., with a larger value of N. The experiments are repeated using Mixer-4 and Mixer-8 for each CFs ranging from $\frac{1}{1024}$ to $\frac{1023}{1024}$ (i.e., $d = 5, N = 4$) and $\frac{1}{512}$ to $\frac{511}{512}$ (i.e., $d = 3, N = 8$), respectively. The average value for each parameter is shown in Table 6.4. It is evident from Table 6.4 that *FloSPA-D* uses minimum mixing steps as guaranteed by *NWayMix* and consumes fewer units of sample compared to *TPG*. Note that with respect to *VOSPA*, sample consumption in *FloSPA-D* is comparable. Furthermore, *FloSPA-D* consumes fewer buffer units, and thus, it outperforms all previous methods in terms of waste generation.

On the same hardware platform, we run all four methods over the full range of *CF*s. Table 6.5 shows the distribution of CPU times needed to generate the mixing graph for target concentrations that lie between $\frac{1}{256}$ and $\frac{255}{256}$ (with $d = 4, N = 4$), $\frac{1}{1024}$ and $\frac{1023}{1024}$ (with $d = 5, N = 4$), and $\frac{1}{512}$ and $\frac{511}{512}$ (with $d = 3, N = 8$). On the average, *FloSPA-D* takes only 0.53 second for ($d = 4, N = 4$), and 4.46 seconds when the accuracy (ϵ) of each target *CF* is further increased, i.e., for $d = 5, N = 4$. Moreover, when a more general mixing model is used with Mixer-8, only 0.34 second of CPU time is required on average to compute the mixing graph for a target *CF* that lies between $\frac{1}{512}$ and $\frac{511}{512}$ (with $d = 3, N = 8$). Note that CPU-time is highly dependent on

[2]*TPG* uses unequal-segmented rotary mixer (Ring-N), whereas *VOSPA, NWayMix*, and *FloSPA-D* use equally segmented rotary mixer (Mixer-N). In our experiments, unequal segments of Ring-N are normalized with respect to the segments in Mixer-N. Additionally, in order to obtain the equivalent mixing models, we use Ring-3, Ring-4 in *TPG* corresponding to Mixer-4 and Mixer-8 used in *VOSPA, NWayMix*, and *FloSPA-D*, respectively.

Table 6.4 Performance of *FloSPA-D*

CF range: $\frac{1}{4^4}$ to $\frac{255}{4^4}$; $(N = 4, d = 4)$				
Method	$\bar{n}_m$	$\bar{n}_w$	$\bar{n}_s$	$\bar{n}_b$
FloSPA-D (Mixer-4)	3.68	3.67	3.35	4.32
NWayMix (Mixer-4)	3.68 (0%)	8.04 (54%)	6.02 (44%)	6.02 (28%)
VOSPA (Mixer-4)	4.13 (11%)	6.70 (45%)	3.34 (-0.3%)	7.36 (41%)
TPG (Ring-3)	4.98 (26%)	7.24 (49%)	3.77 (11%)	7.47 (42%)
CF range: $\frac{1}{4^5}$ to $\frac{1023}{4^5}$; $(N = 4, d = 5)$				
Method	$\bar{n}_m$	$\bar{n}_w$	$\bar{n}_s$	$\bar{n}_b$
FloSPA-D (Mixer-4)	4.67	4.41	3.45	4.96
NWayMix (Mixer-4)	4.67 (0%)	11.01 (60%)	7.51 (54%)	7.51 (34%)
VOSPA (Mixer-4)	5.38 (13%)	8.98 (51%)	3.47 (1%)	9.51 (48%)
TPG (Ring-3)	6.49 (28%)	9.96 (56%)	3.97 (13%)	9.99 (50%)
CF range: $\frac{1}{8^3}$ to $\frac{511}{8^3}$; $(N = 8, d = 3)$				
Method	$\bar{n}_m$	$\bar{n}_w$	$\bar{n}_s$	$\bar{n}_b$
FloSPA-D (Mixer-8)	2.86	5.73	5.64	8.08
NWayMix (Mixer-8)	2.86 (0%)	13.04 (56%)	10.52 (46%)	10.52 (23%)
VOSPA (Mixer-8)	3.26 (12 %)	11.26 (49%)	5.53 (-1%)	13.73 (41%)
TPG (Ring-4)	4.12 (31%)	10.80 (47%)	6.66 (15%)	12.15 (33%)

%-improvement $= \frac{X - \text{Proposed}}{X} \times 100$, $X =$ *NwayMix, VOSPA, TPG*

the value of d. For practical bioassay applications, a high range of accuracy in *CF* can be achieved for $4 \leq N \leq 8$ and $3 \leq d \leq 5$.

6.7.2 Performance Evaluation for Reagent-Saving Mixing

We evaluate the proposed mixture preparation technique extensively on randomly generated synthetic test cases. We choose two datasets with different number of input reagents (Dataset 1 (Dataset 2) that consists of three (four) input reagents) and each data set contains 100 randomly generated target ratios. We run *genMixing*, *FloSPA-M*, and *FloSPA-EM* for all target ratios with two different mixing models supported by Mixer-4 and Mixer-8. Table 6.6 and Table 6.7 show comparative results where savings for different

Table 6.5 CPU-time needed by *FloSPA-D*

CPU time (t in seconds)	Number of Targets $d=4, N=4$	$d=5, N=4$	$d=3, N=8$
$0 \leq t < 1$	231	436	483
$1 \leq t < 2$	12	85	15
$2 \leq t < 3$	6	87	4
$3 \leq t < 4$	1	91	5
$4 \leq t < 5$	2	69	0
$5 \leq t < 6$	0	51	2
$6 \leq t < 7$	0	48	0
$7 \leq t < 10$	0	52	1
$10 \leq t < 20$	2	61	0
$20 \leq t < 30$	1	25	1
$30 \leq t < 50$	0	9	0
$50 \leq t < 76$	0	7	0
$98 \leq t < 126$	0	2	0
Total #target CFs	255	1023	511
Avg. running time	0.53 sec.	4.46 sec.	0.34 sec.

parameters in *FloSPA-M* and *FloSPA-EM* are calculated with respect to the average value of those for *genMixing*, and the CPU-time needed. Note that the running time for *genMixing* is negligibly small in most of the cases and can be ignored. Additionally, in Table 6.6 and Table 6.7, the standard deviation is shown using parenthesis along with the average value of each parameter.

In the case of Dataset 1, the depth of the mixing tree (d) is kept fixed, and hence the accuracy in *CF* is increased in Mixer-8 compared to Mixer-4. The mean and standard deviation of different parameters are listed in Table 6.6.Note that the running time for Mixer-8 increases compared to Mixer-4 as the former provides higher accuracy in *CF*. In the case of Dataset 2, the error tolerance (ϵ) is kept invariant, and hence, the depth of the mixing tree corresponding to Mixer-8 decreases compared to Mixer-4. Table 6.7 shows the mean and standard deviation of different parameters. Compared to Mixer-4, the running time for Mixer-8 reduces because of the smaller value of d. It is evident from simulation results that *FloSPA-M* and *FloSPA-EM* algorithms save valuable reagents significantly compared to *genMixing*. Note that *FloSPA-EM* reduces the number of mixing operations, which, in turn, further reduces the amount of input reagents. However, it takes larger CPU-time compared to *FloSPA-M*, as the search space increases.

Table 6.6 Performance evaluation of mixing algorithms for fixed depth ($d = 3$) of the mixing tree

Dataset 1: # input reagents = 3, # test cases = 100; Mixing model: Mixer-4																			
genMixing ($\epsilon = 0.007$)						*FloSPA-M* ($\epsilon = 0.007$)							*FloSPA-EM* ($\epsilon = 0.007$)						
n_m	n_w	n_{r_1}	n_{r_2}	n_{r_3}	n_r	n_m	n_w	n_{r_1}	n_{r_2}	n_{r_3}	n_r	Time (in sec.)	n_m	n_w	n_{r_1}	n_{r_2}	n_{r_3}	n_r	Time (in sec.)
3.87	8.61	4.9	3.95	3.76	12.61	3.51	5.12	3.97	2.78	2.37	9.12	0.3	3.19	3.81	3.69	2.36	1.76	7.81	0.45
(0.75)	(2.24)	(1.55)	(1.07)	(1.07)	(2.24)	(0.52)	(1.32)	(1.12)	(1.01)	(0.98)	(1.32)	(0.2)	(0.38)	(0.42)	(1.34)	(1.02)	(0.82)	(0.81)	(1.34)
% savings w.r.t *genMixing*						9%	40%	19%	30%	37%	28%	-	18%	56%	25%	40%	53%	38%	-
Dataset 1: # input reagents = 3, # test cases = 100; Mixing model: Mixer-8																			
genMixing ($\epsilon = 0.0009$)						*FloSPA-M* ($\epsilon = 0.0009$)							*FloSPA-EM* ($\epsilon = 0.0009$)						
n_m	n_w	n_{r_1}	n_{r_2}	n_{r_3}	n_r	n_m	n_w	n_{r_1}	n_{r_2}	n_{r_3}	n_r	Time (in sec.)	n_m	n_w	n_{r_1}	n_{r_2}	n_{r_3}	n_r	Time (in sec.)
3.8	19.59	10.95	9.07	7.57	27.59	3.25	9.04	8.01	5.37	3.66	17.04	3.45	2.99	6.65	7.12	4.51	3.02	14.65	5.89
(0.77)	(5.36)	(2.97)	(2.96)	(3.2)	(5.36)	(0.5)	(2.65)	(2.2)	(2.14)	(1.83)	(2.65)	(5.3)	(0.22)	(2.02)	(1.49)	(1.62)	(1.21)	(2.02)	(5.81)
% savings w.r.t *genMixing*						14%	54%	27%	41%	52%	38%	-	21%	66%	35%	50%	60%	47%	-

n_m: number of mixing, n_w: number of waste segments, $n_{r_1}, n_{r_2}, n_{r_3}$: number of segments filled up with reagents R_1, R_2, and R_3, respectively, n_r: Total number of segments filled up with input reagents.

Table 6.7 Performance evaluation for proposed mixing algorithms for fixed error tolerance limit ($\epsilon = 0.001$)

Dataset 2: # input reagents = 4, # test cases = 100; Mixing model: Mixer-4																						
genMixing ($d = 4$)							*FloSPA-M* ($d = 4$)								*FloSPA-EM* ($d = 4$)							
n_m	n_w	n_{r_1}	n_{r_2}	n_{r_3}	n_{r_4}	n_r	n_m	n_w	n_{r_1}	n_{r_2}	n_{r_3}	n_{r_4}	n_r	Time[†]	n_m	n_w	n_{r_1}	n_{r_2}	n_{r_3}	n_{r_4}	n_r	Time[†]
5.52	13.55	6.06	5.15	4.28	2.05	17.55	5.12	9.58	5.22	3.82	3.20	1.34	13.58	6.30	4.84	8.01	4.88	3.33	2.60	1.20	12.01	20.72
(0.97)	(2.90)	(2.02)	(1.81)	(1.66)	(0.76)	(2.90)	(0.90)	(1.61)	(1.53)	(1.33)	(1.46)	(0.50)	(1.61)	(20.13)	(0.56)	(1.07)	(1.30)	(1.16)	(1.09)	(0.43)	(1.07)	(19.05)
% savings w.r.t *genMixing*							7%	29%	14%	26%	25%	35%	23%	-	12%	41%	19%	35%	39%	41%	32%	-
Dataset 2: # input reagents = 4, # test cases = 100; Mixing model: Mixer-8																						
genMixing ($d = 3$)							*FloSPA-M* ($d = 3$)								*FloSPA-EM* ($d = 3$)							
n_m	n_w	n_{r_1}	n_{r_2}	n_{r_3}	n_{r_4}	n_r	n_m	n_w	n_{r_1}	n_{r_2}	n_{r_3}	n_{r_4}	n_r	Time[†]	n_m	n_w	n_{r_1}	n_{r_2}	n_{r_3}	n_{r_4}	n_r	Time[†]
4.03	21.21	10.27	8.15	6.93	3.86	29.21	3.27	11.29	8.51	5.3	4.08	1.4	19.29	2.74	3.42	10.49	8.54	5.35	3.37	1.23	18.49	6.94
(0.74)	(5.21)	(3.28)	(3.38)	(2.95)	(1.87)	(5.21)	(0.47)	(1.9)	(2.3)	(2.13)	(2.1)	(0.62)	(1.9)	(3.05)	(0.51)	(1.89)	(2.27)	(2.08)	(1.87)	(0.42)	(1.89)	(7.14)
% savings w.r.t *genMixing*							19%	47%	17%	35%	41%	64%	34%	-	15%	50%	17%	34%	51%	68%	37%	-

n_m: number of mixing, n_w: number of waste segments, $n_{r_1}, n_{r_2}, n_{r_3}, n_{r_4}$: number of segments filled up with reagents R_1, R_2, R_3, and R_4, respectively, n_r: Total number of segments filled up with input reagents.
[†] Time is measured in seconds.

6.7.3 Performance of *FloSPA* on Real-life Dilution and Mixing Ratios

We consider several real-life test cases where some specific dilution and mixing ratios are used. For example, a sample of 70% ethanol is required in Glucose Tolerance Test in mice, and in E.coli Genomic DNA Extraction [bio]. Also, 95% ethanol is required for total RNA extraction from worms. Assuming $\epsilon = 0.001$ and $N = 4$, 70% and 90% ethanol can be approximated as $\frac{179}{4^4}$ and $\frac{243}{4^4}$, respectively. Hence, the depth of dilution tree $d = 4$. A sample of 10% ($\frac{26}{4^4}$) Fetal Bovine Serum (FBS) is required for in vitro culture of human Peripheral Blood Mononuclear Cells (PBMCs) [bio]. A mixture of three reagents is required in "One step miniprep method" [RCK+15a] and also for the preparation of plasmid DNA by alkaline lysis with SDS-minipreparation [HHC14]. The corresponding mixing ratios are: $\{R_1 : R_2 : R_3 = 50\% : 48\% : 2\%\}$ (approximated as $\{128 : 122 : 6\}$ for $d = N = 4$) and $\{R_1 : R_2 : R_3 = 22\% : 44\% : 34\%\}$ (approximated as $\{56 : 113 : 87\}$ for $N = d = 4$), respectively. Moreover, a target ratio $\{R_1 : R_2 : R_3 : R_4 : R_5 = 40\% : 10\% : 1\% : 1\% : 48\%\}$ is used in "Splinkerette PCR method" [RCK+15a]. Given $\epsilon = 0.001$, the target ratio can be approximated as $\{102 : 26 : 2 : 2 : 124\}$ for Mixer-4 by choosing $d = 4$. Table 6.8 shows the performance of *FloSPA* on these real-life examples compared to other methods.

Table 6.8 Performance of *FloSPA* for real-life test cases

Real-life Dilution: $\epsilon = 0.001, d = 4$

CF	*NWayMix* (Mixer-4)				*FloSPA-D* (Mixer-4)					*VOSPA* (Mixer-4)				*TPG* (Ring-3)			
	n_m	n_w	n_s	n_b	n_m	n_w	n_s	n_b	Time	n_m	n_w	n_s	n_b	n_m	n_w	n_s	n_b
$\frac{26}{4^4}$	4	9	5	8	4	3	1	6	0.00 sec.	4	5	1	8	6	8	1	11
$\frac{29}{4^4}$	4	9	5	8	4	5	1	8	0.09 sec.	5	7	1	10	6	7	1	10
$\frac{179}{4^4}$	4	9	8	5	4	3	5	2	1.67 sec.	5	7	4	7	5	6	4	6
$\frac{243}{4^4}$	4	9	9	4	4	4	7	1	0.51 sec.	4	7	7	4	4	7	7	4

Real-life mixture preparation: $\epsilon = 0.001, d = 4$; Mixing model: Mixer-4

Ratio	*genMixing*			*FloSPA-M*				*FloSPA-EM*			
	n_m	n_w	n_r	n_m	n_w	n_r	Time	n_m	n_w	n_r	Time
$\{56 : 113 : 87\}$	5	12	16	5	10	14	4.56 sec.	4	5	9	10.03 sec.
$\{128 : 122 : 6\}$	4	9	13	4	6	10	0.25 sec.	4	6	10	1.00 sec.
$\{102 : 26 : 2 : 2 : 124\}$	7	18	22	6	10	14	8.25 sec.	6	8	12	30.09 sec.

n_m: number of mixing, n_w: number of waste segments, $n_s(n_b)$: number of segments filled up with sample (buffer), n_r: Total number of segments filled up with input reagents.

6.8 Conclusions

In this chapter, we have presented the problem of automated sample preparation using flow-based microfluidic biochips and proposed a dilution algorithm that minimizes reagent usage under the constraint of deploying minimum mixing steps. Additionally, two algorithms for reagent-saving mixture-preparation using flow-based biochips have been developed. We use an N-segment rotary mixer for implementing a generalized mixing model. The proposed algorithms utilize an SMT-based solving engine that is capable of handling both dilution and mixing problems within the same framework. Although for mixture preparation, we have assumed the same cost for each of the input reagents, the proposed modeling can be modified for finding solutions when different cost metrics are assigned to them.

7

Storage-Aware Algorithms for Dilution and Mixture Preparation with Flow-Based Lab-on-Chip

Microfluidic biochips have been considered as one of the key technologies for automatic sample preparation that includes dilution and mixing of fluids in certain ratios. Dilution and mixing are implemented on biochips through a sequence of basic fluid-mixing and splitting operations. These steps are abstracted using a mixing graph. During this process, on-chip storage units are needed to store intermediate fluids to be used later in the sequence. This allows to optimize the reactant costs, to reduce the sample-preparation time, and/or to achieve the desired ratio. Several approaches have been proposed for sample preparation on LoCs (see e.g., [TUTA08, RBC10, LCH15, HLH15, LSH15, BPR$^+$16]) considering different optimization objectives such as minimization of the (1) number of mixing operations (i.e., time), (2) consumption of valuable reagents, and (3) amount of waste generation.

Most of the existing LoC sample preparation methods indeed consider these three optimization objectives, but they are oblivious of a crucial fact: the number of storage units available on a CFMB is usually limited. A storage unit is needed when an intermediate fluid-mixture needs to be stored for subsequent use. Unfortunately, maintaining an on-chip storage module in a CFMB is significantly expensive [TUTA08, MQ07]. They do not only increase the area of the chip and require additional channel connections, but also lead to a more complicated fabrication, calibration, and testing. As a consequence, corresponding sample preparation methods need to consider this restriction. Moreover, in cases where sample preparation is possible without any intermediate storage unit, the algorithm should exploit the available storage units at the platform in order to further reduce the consumption of reagents.

In this chapter, we discuss a storage-aware sample preparation method for mixing two or more biochemical reagents on a CFMB which addresses these issues. To this end, an SAT-based approach is presented, which allows to efficiently check several options of generating the desired target ratio and, eventually, choosing the one which makes the best use of the available storage units while, at the same time, optimizing sample preparation costs and/or time. Implicitly, the proposed method also guarantees that no solution is chosen, which requires more storage units than available for the given platform and flags when a solution does not exist.

7.1 Related Works

Liu *et al.* proposed a tree pruning and grafting method (called *TPG* [LSH15]) that starts from an initial mixing tree (based on a (1:1) mixing model) and transforms it for obtaining a dilution graph for unequally segmented rotary mixer (Ring-N). In [HLH15], a volume-oriented sample preparation algorithm (called *VOSPA*) has been introduced, which employs a greedy strategy. Lei *et al.* [LLH16] proposed a network-flow-based multi-objective dilution method that utilizes the full flexibility of the multiple mixing model offered by Mixer-N. Later, a flow-based sample preparation algorithm (called *FloSPA*) was proposed in [BPR$^+$16] that can handle dilution and mixing within one framework and fully utilize the power of the multiple mixing model supported by the Mixer-N. A summary of these existing CFMB-based sample-preparation methods is provided in Table 7.1.

Table 7.1 Summary of CFMB sample preparation algorithms

Method	#-Input Reagents	Uses all Possible Mixing Ratios of Underlying Mixing Model?	Considers Number of Storage-Units?
NWayMix [BPR$^+$16]	2	No	No*
TPG [LSH15]	2	No	No$^\diamond$
VOSPA [HLH15]	2	No	No$^\diamond$
Flow-based [LLH16]	2	Yes†	No$^\diamond$
FloSPA [BPR$^+$16]	≥ 2	Yes	No$^\diamond$
Proposed	≥ 2	Yes	Yes

*Does not utilize any storage unit at all (and, hence, yields rather expensive solutions when storage units are available).

$^\diamond$Provides an invalid solution when the number of storage units is insufficient compared to what is necessitated by the algorithm.

†Is computationally expensive when the number of segments in Mixer-N increases.

7.2 Storage-Aware Sample Preparation

In this section, the proposed method is described in detail. Before going into detailed description, let us consider a motivating example.

Example 7.2.1 Suppose we need to prepare a mixing ratio {sample : buffer = 125 : 131} on a CFMB platform that supports only two on-chip storage units. The mixing graph determined with existing sample preparation methods, e.g., *VOSPA* [HLH15] and *FloSPA* [BPR+16], require four and five storage units as shown in Figure 7.1(a) and Figure 7.1(b), respectively. Hence, these results obtained by these approaches are useless. Moreover, since a dilution problem is considered here, a mixing graph requiring zero storage unit as shown in Figure 7.1(c) can be determined using the *NWayMix* [BPR+16] approach. But since this does not utilize the available storage units, a total of nine units of the sample are required in this case (cf. Figure 7.1(c)) – a very expensive solution. In contrast, the desired mixing ratio can be realized more efficiently as shown in Figure 7.1(d). The improved solution not only requires no more than the available number of storage units (hence, it is a valid solution) but also exploits them to reduce the total number of sample-units from 9 to 4.

The above-mentioned example motivates a storage-aware sample-preparation method, which does not generate a mixing graph exceeding the number of available storage units and, at the same time, fully exploits them in order to reduce the costs.

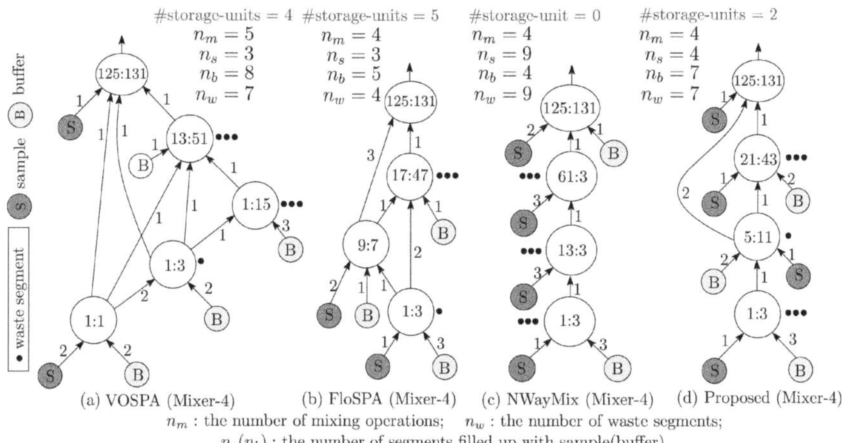

Figure 7.1 Dilution graph generated with (a) *NWayMix*, (b) *VOSPA*, (c) *FloSPA*, and (d) the proposed method for {sample:buffer = 125:131}.

7.2.1 Overview

The main idea is to utilize mixing graphs generated by earlier approaches as basis, which already provide an option how to eventually realize the desired concentration ratio [HLH15, BPR+16]. Next, the mixing tree is augmented with additional nodes (allowing to use further input reagents) and edges (allowing to share intermediate fluids) – eventually providing several further options for realizing the desired input ratio. However, in order to determine the one which gives the minimum reagent usage and, at the same time, satisfies the limitations in storage units is a computationally complex task. In order to cope with this complexity, we use the computational power of Boolean satisfiability solvers [dMB08, BPF15], which already have been found effective for similar tasks in the design of LoCs (see e.g., [KWHD14, GWY+17]). The main idea is to symbolically represent all possible options (given by the augmented mixing graph) and to extend this representation by constraints enforcing the storage limitation. Finally, the resulting formulation is passed to a solving engine, which either determines a satisfying solution (out of which a mixing graph satisfying the storage constraints can be derived) or proves that, considering the available options, no such solution exists.

In the following, the proposed approach is described in two steps. First, we consider dilution problems only, i.e., the case where only two fluids (a sample and a buffer) are mixed. Here, in fact, every ratio can be realized with zero storage unit (although the storage units which are available anyway can be utilized to reduce the costs of the sample preparation). Afterwards, the general case of mixing is covered, i.e., the case where more than two fluids are mixed. This requires stricter constraints to be satisfied and, hence, is described in a separate subsection. The main steps of both flows for dilution and mixing are summarized in Figures 7.2 and 7.3, respectively.

7.2.2 Storage-Aware Dilution

We describe the proposed method for dilution. Given a desired ratio of sample and buffer, the method starts with a mixing graph generated with an existing sample preparation method. Here, the approach called *NWayMix* and proposed in [BPR+16] is suitable due to two main reasons[1]: First, *NWayMix* generates a target ratio using Mixer-N with a minimum number of mixing steps i.e., a minimum sample preparation time. Second, the mixing tree

[1]Nevertheless, the proposed method can similarly be applied using other sample-preparation methods.

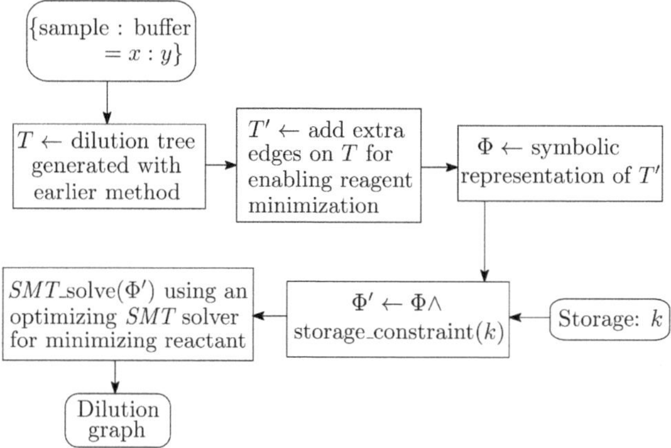

Figure 7.2 Flowchart of the dilution algorithm.

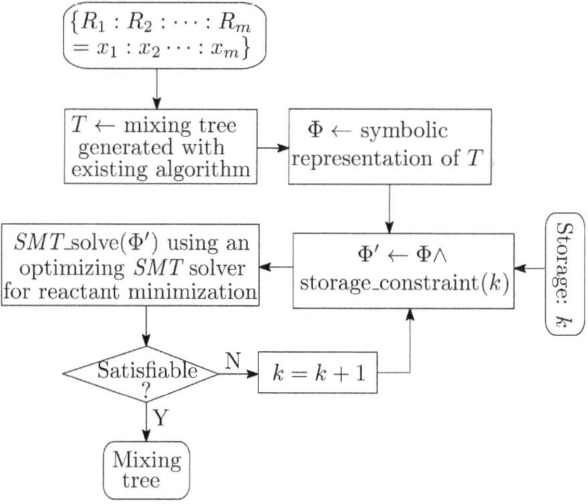

Figure 7.3 Flowchart of the mixing algorithm.

generated by *NWayMix* resembles a chain (i.e., a skewed graph) and, hence, can be executed on a single CFMB-mixer without any on-chip storage unit for intermediate fluids. Figure 7.4(a) sketches the resulting graph.

The main idea is to augment the mixing graph produced by *NWayMix* with additional leaf-nodes (input reagents) and edges – allowing for further options to realize the desired ratio. This is sketched by means of blue leaf-nodes and edges in the graph shown in Figure 7.4(b). The general structure of such transformation is shown in Figure 7.4(c). They eventually represent further

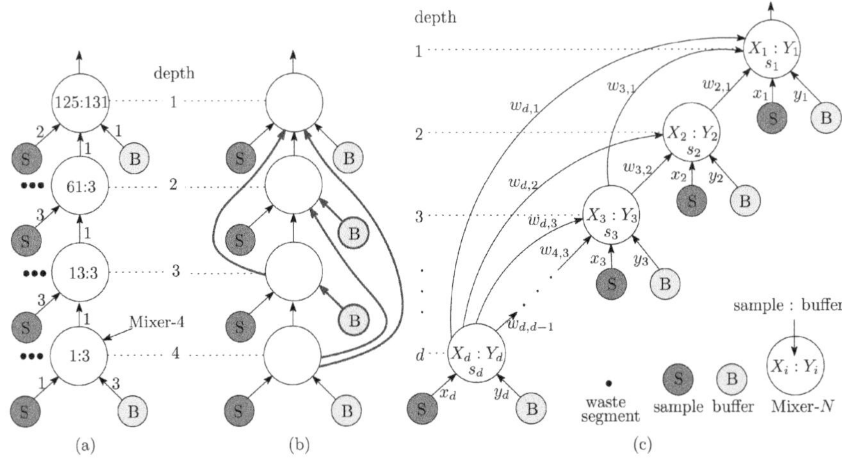

Figure 7.4 (a) Dilution tree generated by *NWayMix* for target ratio {sample:buffer = 125:131}, (b) dilution graph after adding extra edges and reagent nodes for enabling reagent minimization, and (c) general structure of dilution graph.

options for mixing in which intermediate results (stored in storage units) are re-used. This yields the question what inputs shall be used in each mixing step (i.e., what edges shall remain in the mixing graph). In order to determine the best possible result, all possibilities should be checked for this purpose. Since doing this enumerately is infeasible, we formulate this problem in terms of a satisfiability instance. To this end, we introduce the following free variables:

Node variables: For each mixing node at depth i in the mixing graph, we define two rational variables X_i and Y_i $(1 \leq i \leq d)$ that denote the ratio between sample and buffer of the resulting mixing operation at depth i.

Reagent variables: The input reagents (sample and buffer) can be used in any mixing node at depth i, where $1 \leq i \leq d$. For denoting the number of segments filled with sample and buffer in a Mixer-N at depth i, two integer variables x_i and y_i are associated, respectively.

Segment-sharing variables: The integer variables $w_{i,j}$ represent the number of segments that are used in Mixer-N at depth j from Mixer-N at depth i, where $1 \leq j < i \leq d$.

Storage variables: An integer variable s_i is associated with each mixing node at depth i $(1 \leq i \leq d)$ for denoting the number of on-chip storage units required for executing the portion of induced subgraph containing mixing nodes at depth j, where $i \leq j \leq d$. Note that s_1 denotes the storage requirement for the entire mixing graph.

The annotations in Figure 7.4(c) provide all variables used in this case. However, passing this representation to a solving engine would yield an arbitrary assignment to all variables and, hence, a mixing tree that realizes arbitrary ratios in each depth using an arbitrary number of storage units – obviously a solution which is neither valid nor desired. Hence, we need to further constrain the introduced variables so that indeed the desired result is determined.

Enforcing the Desired Ratio

First, the correctness of the respective mixing ratios is enforced. To this end, the following constraints are introduced for each mixing node at depth i of the mixing graph:

$$x_i + w_{d,i}X_d + w_{d-1,i}X_{d-1} + \cdots + w_{i+1,i}X_{i+1} = NX_i \quad (7.1)$$

$$y_i + w_{d,i}Y_d + w_{d-1,i}Y_{d-1} + \cdots + w_{i+1,i}Y_{i+1} = NY_i \quad (7.2)$$

Note that each mixing node at depth i can fill its segments with sample, buffer, or any unused fluid segments produced at depth $j > i$. Hence, the desired ratio of sample and buffer at depth i i.e., $\{X_i : Y_i\}$ is determined with these equations. Furthermore, the non-linearity of above equations can be removed easily by adding few extra constraints as carried out in [BPR$^+$16]. This transformation helps to run powerful sound and complete SMT-solvers [dMB08, BPF15] and speed up the computation significantly. Additionally, we need to ensure that all N input segments for a Mixer-N must be filled with intermediate fluids or reagents, whereas Mixer-N can serve at most N segments to other mixers. The required consistency constraints at depth i are enforced by:

$$x_i + y_i + w_{d,i} + w_{d-1,i} + \cdots + w_{i-1,i} = N \quad (7.3)$$

$$w_{i,i-1} + w_{i,i-2} + \cdots + w_{i,i} \leq N \quad (7.4)$$

Besides that, all weights must satisfy $0 \leq w_{i,j} \leq N - 1$, for $1 \leq j < i \leq d$. Analogously, $0 \leq x_i, y_i \leq N - 1$ for $1 \leq i \leq d$. Finally, the constraint $(X_1 = x) \wedge (Y_1 = y)$ guarantees that the desired target ratio of sample and buffer $\{x : y\}$ is produced.

Enforcing the Available Number of Storage-Units

Next, we have to enforce that not more than the available number of storage units is used. To this end, we compute the number of required storage units, which would be needed according to a particular assignment of the variables by traversing the mixing nodes in a bottom-up fashion. Recall that, for each depth, a storage variable s_i is available, which denotes the number of on-chip storage units required for executing the portion of induced subgraph containing mixing nodes at depth j, where $i \leq j \leq d$. This amount is determined by the following equation:

$$s_i = \begin{cases} s_{i+1} + w_{d,i} + w_{d-1,i} + \cdots + w_{i+2,i}, \; 1 \leq i < d \\ 0, \quad i = d \end{cases} \tag{7.5}$$

Accordingly, s_1 denotes the storage requirement for the entire mixing graph. Restricting this variable to the number k of available storage units, i.e., enforcing $s_1 \leq k$, only allows assignments for all other variables which eventually represent solutions that do not use more than k storage units.

Figure 7.5 shows the induced subgraph of the general mixing graph from Figure 7.4(b) used in the storage calculation. Note that no storage unit is required for subgraph containing mixing node at depth d only. Hence, $s_d = 0$. However, for computing storage requirement of the induced subgraph given in Figure 7.5, we need to add s_{i+1} and the total number of on-chip storage units used for storing unused segments at depth $i+2, i+3, \ldots, d$, which are used in the mixing node at depth i. Hence, $s_i = s_{i+1} + w_{d,i} + w_{d-1,i} + \ldots + w_{i+2,i}$. It can be easily verified that the bottom-up computation of s_1 gives the minimum number of on-chip storage units for executing a dilution graph on the CFMB platform equipped with a single mixer. Note that it is up to the optimizing SMT-solver [dMB08, BPF15] that finds the reagent minimal solution on the general dilution tree (Figure 7.4(b)) that gives minimal reagent solution satisfying storage constraint.

Example 7.2.2 Figure 7.6 shows solutions realizing the target ratio {sample: buffer = 125:131} with the least reactant-cost based on the graph of Figure 7.4(a) and considering the availability of a different number of on-chip storage units on the target architecture. The dilution algorithm modifies the graph based on the number of available storage units (#storage units = $0, 1, \ldots, 5$).

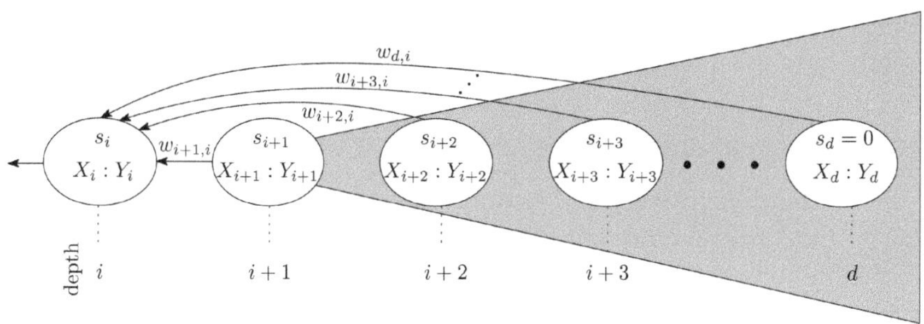

Figure 7.5 Structure of the induced subgraph in the general dilution tree, containing mixing nodes at depth $i, i + 1, \ldots, d$, used for storage computation.

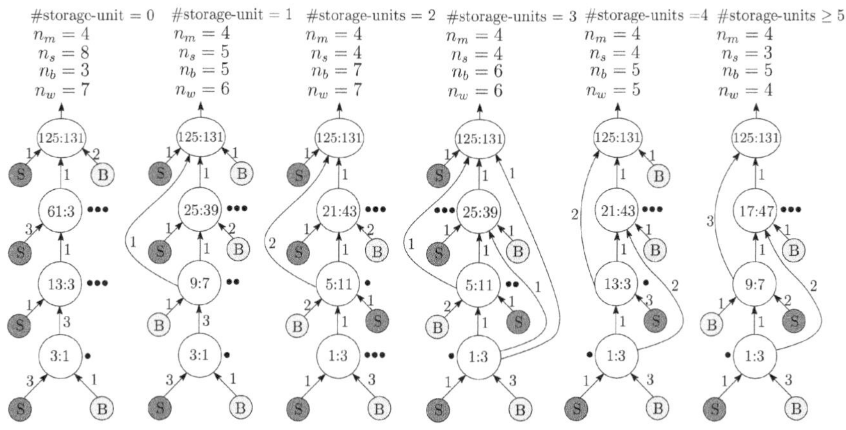

Figure 7.6 Reagent minimal solution for the target ratio {sample:buffer $= 125 : 131$} considering the availability of different number of available on-chip storage units on the target CFMB architecture.

7.2.3 Overview of the Storage-Aware Mixing

Mixing algorithm basically follows the same idea as the method for dilution. However, in contrast to dilution, it is not always possible to perform mixing of three or more reagents with zero on-chip storage unit. This is due to the inherent tree structure of the mixing graph, which commonly appears in mixture preparation. As before, we start with a basis mixing graph produced by previously proposed methods (here, e.g., *genMixing* [BPR+16]) and introduce additional variables and constraints to provide a symbolic representation of all possible options out of which the best one satisfying the storage limitation is determined by the solving engine. To this end, we particularly have to adjust the storage constraints[2].

Enforcing the Available Number of Storage Units

Figure 7.7 sketches a generic node in a mixing graph as well as a simplified notation of the variables that are used in storage calculation. On a mixing tree, a mixing node may take a segment of fluid from one of its subtrees or it may take input reagents to fill one of its N segments. Hence, there can be at most N subtrees possible for a mixing node. In Figure 7.7, w_i (segment-sharing variable) and r_i (reagent variable) denote the number of segments of a Mixer-N filled with fluids taken from the root node of subtree-i and the input reagent R_i, respectively. Besides that, s_i denotes again the storage requirement of subtree-i. Then, the number of storage units s can be determined by the following equation:

$$s = \bigvee_{j=1}^{N} \left(\sum_{i=1}^{j-1} w_i + s_j + \sum_{i=j+1}^{N} w_i \right) \tag{7.6}$$

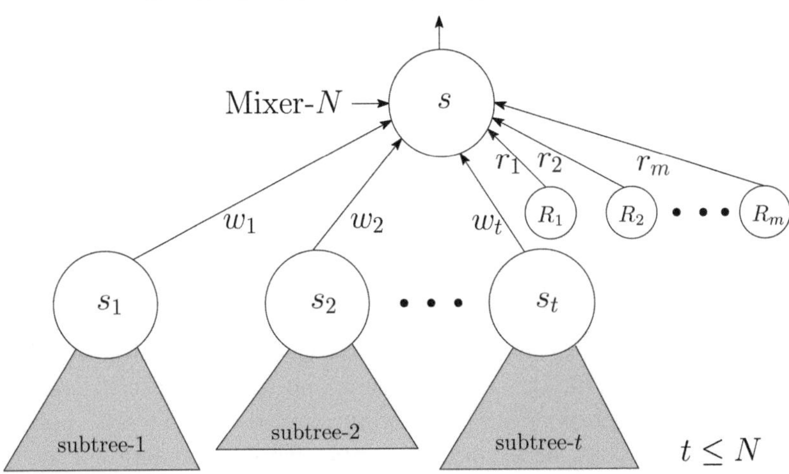

Figure 7.7 Structure of a node in the mixing tree used in storage computation.

[2]We are not repeating the definition of the variables and constraints enforcing the target ratios. They are basically identical to the ones used for dilution.

Note that Equation 7.6 symbolically represents the various scheduling issues for each node. As in dilution, enforcing the corresponding s-variables of all mixing nodes in the mixing tree to be smaller than or equal to the number k of available storage units will eventually only allow solutions which can be realized under this restriction.

Example 7.2.3 Figures 7.8(a)–(c) show the mixing graphs for the target ratio $\{R_1 : R_2 : R_3 : R_4 = 22 : 14 : 14 : 14\}$ obtained by *genMixing* [BPR+16], *FloSPA* [BPR+16] and by the proposed method, respectively. Note that Figure 7.8(a) requires two storage units. Figure 7.8(b) shows a

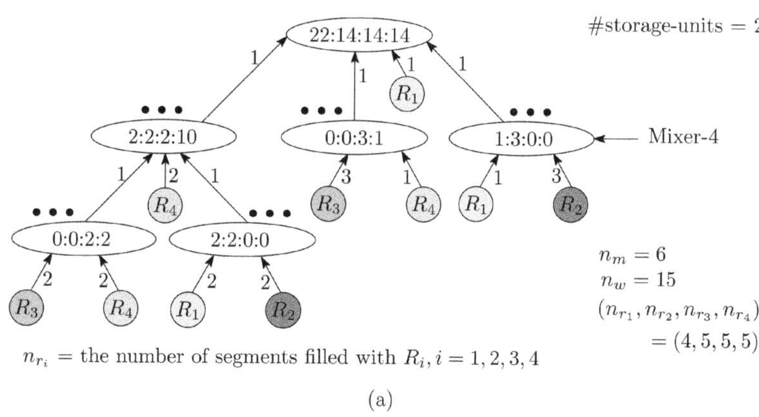

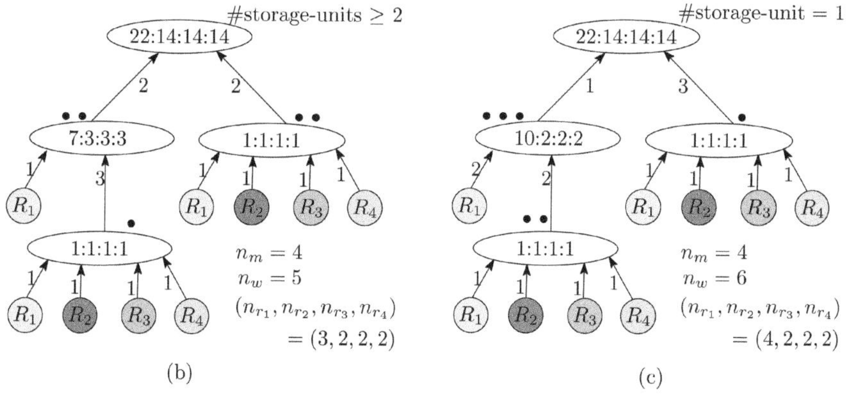

Figure 7.8 Mixing tree for the target ratio $\{R_1 : R_2 : R_3 : R_4 = 22 : 14 : 14 : 14\}$ generated by (a) *genMixing* [BPR+16], (b) *FloSPA* [BPR+16] and the proposed method for #storage-units ≥ 2, and (c) the storage-aware mixing algorithm when #storage-unit = 1.

cheaper solution obtained by *FloSPA* using two storage units. On the other hand, the proposed method ensures that no solution exists with zero storage unit starting from Figure 7.8(a); it also provides a solution with one storage unit (Figure 7.8(b)), and with two storage units (Figure 7.8(c)).

7.3 Experimental Results

The methods described above are compared with several state-of-the-art sample-preparation methods, namely *NWayMix* [BPR+16] and *VOSPA* [HLH15] for dilution as well as with *FloSPA* [BPR+16] for dilution and general mixing. In the following, the obtained results for both cases are summarized.

7.3.1 Performance for Dilution

VOSPA [HLH15] and *FloSPA* [BPR+16] focus on reagent minimization and do not take the number of available storage units into account. This can have crucial consequences as illustrated by a first series of experiments summarized in Figure 7.9. Here, we have generated dilution graphs for all possible target ratios of sample and buffer by varying $d = 4, 5$ for $N = 4$ and listed for how many ratios a particular number of storage units is required. As can be seen, for the vast majority of ratios, both approaches require a substantial amount of storage units. If this amount is not available on the considered platform, the obtained result is useless. In contrast, the storage-aware sample

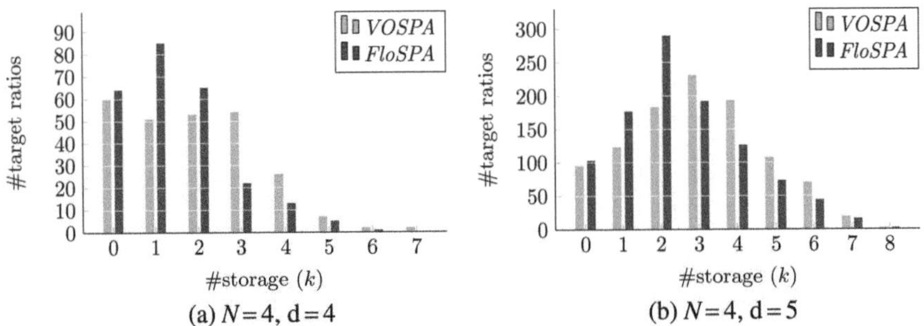

Figure 7.9 Histogram of the number of storage units required by VOSPA and FloSPA for all target ratios corresponding to (a) {sample:buffer = x : $256 - x$}, where $1 \leq x \leq 255$, i.e., $N = 4, d = 4$ and (b) {sample:buffer = x : $1024 - x$}, where $1 \leq x \leq 1023$, i.e., $N = 4, d = 5$.

Table 7.2 Performance of proposed dilution algorithm.

Storage-Aware Approach					*NWayMix* [BPR+16]	*VOSPA* [HLH15]	*FloSPA* [BPR+16]
k	$\bar{n}_m$	$\bar{n}_w$	$\bar{n}_s$	$\bar{n}_b$	#Storage-Unit = 0	#Storage-Units = 7	#Storage-Units = 6
0	3.68	5.39	4.22	5.18	$\bar{n}_m = 3.68$	$\bar{n}_m = 4.13$	$\bar{n}_m = 3.68$
1	3.68	4.19	3.51	4.67	$\bar{n}_w = 8.04$	$\bar{n}_w = 6.70$	$\bar{n}_w = 3.66$
2	3.68	3.80	3.40	4.40	$\bar{n}_s = 6.02$	$\bar{n}_s = 3.34$	$\bar{n}_s = 3.36$
3	3.68	3.71	3.38	4.32	$\bar{n}_b = 6.02$	$\bar{n}_b = 7.36$	$\bar{n}_b = 4.30$
≥ 4	3.68	3.66	3.36	4.30			

preparation algorithm is capable of determining a mixing graph for all ratios and for all given numbers of available storage units (even zero). This is a clear improvement compared to *VOSPA* and *FloSPA* since the desired dilution can always be realized using the method proposed in this work.

Moreover, even with respect to costs, significant improvements can be observed as shown in Table 7.2. Here, we have generated dilution graphs for all possible target ratios in $\{\text{sample} : \text{buffer} = x : 256 - x\}$, where $1 \leq x \leq 255$, i.e., $N = 4, d = 4$ and listed the average number of mixing steps ($\bar{n}_m$), waste segments ($\bar{n}_w$), and number of segments filled with sample ($\bar{n}_s$) and buffer ($\bar{n}_b$) for a different number of available on-chip storage units (k) in case of the storage-aware dilution algorithm. For the previously proposed approaches, we list the best results, i.e., obtained with zero storage unit in case of *NWayMix*, obtained with seven storage units in case of *VOSPA*, six storage units in case of *FloSPA*.

Several issues can be observed here: First, *NWayMix* always uses zero storage unit. By this, other potential solutions are missed out since a few storage units are usually available on any platform which, as shown by the numbers in Table 7.2, can be exploited to improve e.g., the number of input reagent units. Either way, even with zero storage unit, the storage-aware approach still determines better results than *NWayMix*. It can also be observed from Table 7.2 that the storage-aware approach only needs two (four) storage units in order to get a comparable performance with respect to *VOSPA* (*FloSPA*) requiring a total of seven (six) storage units.

7.3.2 Performance for Mixing

Finally, we have experimented with several actual mixing ratios and the number of on-chip storage units (k) as summarized in Table 7.3. Here, we list the considered mixing ratios as well as the number of mixing steps (n_m), waste segments (n_w), and the total number of segments filled with input reagents (n_r). It can be observed that the stoarage-aware mixing algorithm

Table 7.3 Performance of the proposed mixing algorithm

#	Mixing Ratio	*genMixing* [BPR+16] n_m n_w n_r	k	Proposed Method $k=0$ (n_m, n_w, n_r)	$k=1$ (n_m, n_w, n_r)	$k \geq 2$ (n_m, n_w, n_r)
1	90:90:76	5 12 16	1	$(4, 8, 12)$	$(5, 6, 10)$	$(5, 6, 10)$
2	20:14:16:14	4 9 13	1	no sol.	$(4, 6, 10)$	$(4, 5, 9)$
3	27:7:13:17	5 12 16	1	no sol.	$(4, 8, 12)$	$(4, 8, 12)$
4	68:86:56:46	6 15 19	1	no sol.	$(5, 9, 13)$	$(6, 9, 13)$
5	27:25:57:69:78	8 21 25	2	no sol.	$(7, 15, 19)$	$(7, 13, 17)$
6	30:24:55:68:79	8 21 25	2	no sol.	$(6, 13, 17)$	$(6, 12, 16)$
7	30:24:155:38:9	8 21 25	2	no sol.	$(7, 15, 19)$	$(7, 13, 17)$
8	300:499:225	7 18 22	1	$(5, 4, 9)$	$(5, 4, 9)$	$(5, 4, 9)$
9	57:28:6:6:6:3:150	9 24 28	2	no sol.	$(6, 10, 14)$	$(6, 10, 14)$
10	102:26:3:3:122	8 21 25	2	no sol.	no sol.	$(7, 12, 16)$
11	4:6:10:14:22:26:174	10 27 31	3	no sol.	no sol.	$(7, 14, 18)$
12	26:15:51:26:5:5:1:127	12 33 37	3	no sol.	no sol.	$(10, 24, 28)$

parameter values corresponding to *FloSPA* [BPR+16] are highlighted in yellow.

can perform mixture preparation using only a few on-chip storage units. In fact, all ratios can be realized with at most two storage units – in many cases, even only one is sufficient. Moreover, it not only takes same or fewer number of on-chip storage units compared to *genMixing* but also produces the target ratio with a samller number of mixing steps, waste, and input segments.

Note that mixture preparation using *FloSPA* reduces the input reagent consumption without considering the limited availability of on-chip storage units. For example, in case of Mixture 3 and Mixture 4 shown in Table 7.3, the mixing graph generated with *FloSPA* may require one/two on-chip storage units as both solutions (for $k = 1$ and $k \geq 2$) fill the same number of segments with input reagents, i.e., both solutions meet the optimization objective of minimum reagent usage. Moreover, for Mixtures 1–7 in Table 7.3, the proposed method can produce the same mixture with a smaller number of storage units by consuming little more input reagents compared to *FloSPA*.

7.4 Conclusions

In this chapter, we discussed a storage-aware sample-preparation method for continuous-flow microfluidic biochips. We show that the augmented mixing graphs determined by previously proposed sample-preparation methods eventually providing several further options for realizing the desired input ratio. Afterwards, Boolean satisfiabiliy solvers are utilized to determine the option that gives the minimum reagent usage and, at the same time, satisfies the limitations in storage elements. This provides significant benefits as it can be ensured that a generated mixing graph indeed can be executed on the biochip device (compared to previously proposed solutions, which may generate mixing graphs that require more storage units than available and, hence, are useless). Moreover, the stoarge-aware sample prepration explicitly allows to fully exploit the available number of storage units, e.g., in order to reduce the use of reagents.

8

Conclusion and Future Directions

In this book, several computer-aided design (CAD) algorithms for sample preparation with microfluidic biochips have been illustrated. We have discussed multiple-target dilution algorithms that can generate a set of target concentrations with digital microfluidic biochips (DMFBs) while optimizing sample-preparation cost and time. When the required concentrations in a target set follow certain gradient patterns such as linear, exponential, and parabolic specially designed algorithms are needed to handle these cases for further optimizations of different parameters. We have also described a number of algorithms that are capable of exploiting the underlying combinatorial properties of the gradient pattern for reducing sample-preparation cost significantly, compared to other general-purpose sample preparation techniques. Next, a concentration-resilient ratio-selection method has been illustrated that selects a suitable mixing ratio among possible ranges so that sample preparation can be accomplished efficiently. The integer linear programming-based method for concentration-resilient mixture preparation terminates very fast and outputs the optimum solution considering both uniform and weighted cost of reagents. We have studied the algorithmic aspects of sample preparation for flow-based microfluidic biochips that support generalized mixing models. We have shown that the satisfiability-based modeling technique is very effective for handling dilution and mixing problems on flow-based microfluidic biochips. Finally, we describe the storage-aware sample preparation algorithms for flow-based microfluidic biochips.

We suggest a number of new problems that can be considered as future research directions in the design automation of microfluidic sample preparation. Most of the existing sample preparation algorithms on DMFB consider (1:1) mixing model. Recent emergence of micro-electrode dot array (MEDA) architectures offer more flexibilities, e.g., variable size droplets, support for multiple mixing models over conventional digital microfluidics in performing various operations. It will be a fruitful area of future research to optimize sample preparation time and cost by considering these flexibilities offered by the MEDA. On the other hand, sample preparation on fully programmable valve array, where variable-size mixers can be easily implemented, can also be explored. One can also explore the algorithmic sample preparation for hybrid rotary mixer architectures. It may also be worthwhile to expand the scope of the flow-based sample preparation algorithms by addressing several error management issues concerning imprecise metering.

Bibliography

[AB14] Ismail Emre Araci and Philip Brisk. Recent developments in microfluidic scale integration. *Current Opinion in Biotechnology*, 25:60–68, 2014.

[AHS+13] Adam R. Abate, Tony Hung, Ralph A. Sperling, Pascaline Mary, Assaf Rotem, Jeremy J. Agresti, Michael A. Weiner, and David A. Weitz. Dna sequence analysis with droplet-based microfluidics. *Lab Chip*, 13:4864–4869, 2013.

[AHU74] Alfred V. Aho, John E. Hopcroft, and Jeffrey D. Ullman. *The Design and Analysis of Computer Algorithms*. Addison-Wesley, 1974.

[And01] Jennifer M. Andrews. Determination of minimum inhibitory concentrations. *Journal of Antimicrobial Chemotherapy*, 48(1):5–16, July 2001.

[ASS+17] V. Agarwal, A. Singla, M. Samiuddin, S. Roy, T.-Y. Ho, I. Sengupta, and B. B. Bhattacharya. Reservoir and mixer constrained scheduling for sample preparation on digital microfluidic biochips. In *Proc. ASP-DAC*, 702–707, 2017.

[BB07] Partha Bhowmick and Bhargab B. Bhattacharya. Fast polygonal approximation of digital curves using relaxed straightness properties. *IEEE Trans. PAMI*, 29(9):1590–1602, 2007.

[BBB14] Sukanta Bhattacharjee, Ansuman Banerjee, and Bhargab B. Bhattacharya. Sample preparation with multiple dilutions on digital microfluidic biochips. *Computers Digital Techniques, IET*, 8(1):49–58, Jan 2014.

[BBH+13a] Sukanta Bhattacharjee, Ansuman Banerjee, Tsung-Yi Ho, Krishnendu Chakrabarty, and Bhargab B. Bhattacharya. Eco-friendly sample preparation with concentration gradient on a digital microfluidic biochip. In *Proc. ICECCS*, 1–11, 2013.

[BBH+13b] Sukanta Bhattacharjee, Ansuman Banerjee, Tsung-Yi Ho, Krishnendu Chakrabarty, and Bhargab B. Bhattacharya. On producing linear dilution gradient of a sample with a digital microfluidic biochip. In *Proc. ISED*, 77–81, 2013.

[BBH+ar] Sukanta Bhattacharjee, Ansuman Banerjee, Tsung-Yi Ho, Krishnendu Chakrabarty, and Bhargab B. Bhattacharya. Efficient generation of dilution gradients with digital microfluidic biochips. *IEEE Trans. on CAD*, 2018 (to appear).

[BGB17] Tapalina Banerjee, Sarmishtha Ghoshal, and Bhargab B. Bhattacharya. COMSOL-based design and validation of dilution algorithm with continuous-flow lab-on-chip. *INAE Letters*, 2(2):55–63, 2017.

[bio] Bio-protocol. http://www.bio-protocol.org/.

[BKE16] Margarita Barbatsi, Michael Koupparis, and Anastasios Economou. A new flow-injection chromatography method exploiting linear-gradient elution for fast quantitative screening of parabens in cosmetics. *Anal. Methods*, 8:8337–8344, 2016.

[BMMG+11] D. Brassard, L. Malic, C. Miville-Godin, F. Normandin, and T. Veres. Advanced EWOD-based digital microfluidic system for multiplexed analysis of biomolecular interactions. In *Proc. MEMS*, 153–156, January 2011.

[Boo51] Andrew D. Booth. A signed binary multiplication technique. *The Quarterly Journal of Mechanics and Applied Mathematics*, 4, 1951.

[BPF15] Nikolaj Bjørner, Anh-Dung Phan, and Lars Fleckenstein. *vZ* - an optimizing SMT solver. In *Proc. of TACAS*, 194–199, 2015.

[BPR+16] Sukanta Bhattacharjee, Sudip Poddar, Sudip Roy, Junin-Dar Huang, and Bhargab B. Bhattacharya. Dilution and mixing algorithms for flow-based microfluidic biochips. *IEEE Trans. on CAD*, 36(4):614–627, 2016.

[BWHB18] Sukanta Bhattacharjee, Robert Wille, Juinn-Dar Huang, and Bhargab B. Bhattacharya. Storage-aware sample preparation using flow-based microfluidic labs-on-chip. In *Proc. of DATE*, 1399–1404, 2018.

[CHDW12] Nate J. Cira, Jack Y. Ho, Megan E. Dueck, and Douglas B. Weibel. A self-loading microfluidic device for determining the minimum inhibitory concentration of antibiotics. *Lab Chip*, 12:1052–1059, 2012.

[CLC+11] Curtis D. Chin, Tassaneewan Laksanasopin, Yuk Kee Cheung, David Steinmiller, Vincent Linder, Hesam Parsa, Jennifer Wang, Hannah Moore, Robert Rouse, Gisele Umviligihozo, Etienne Karita, Lambert Mwambarangwe, Sarah L Braunstein, Janneke van de Wijgert, Ruben Sahabo, Jessica E Justman, Wafaa El-Sadr, and Samuel K Sia.

Microfluidics-based diagnostics of infectious diseases in the developing world. *Nature Medicine*, 17:1015–1019, 2011.

[CLH13] Ting-Wei Chiang, Chia-Hung Liu, and Juinn-Dar Huang. Graph-based optimal reactant minimization for sample preparation on digital microfluidic biochips. In *Proc. VLSI-DAT*, 1–4, 2013.

[CLS12] Curtis D. Chin, Vincent Linder, and Samuel K. Sia. Commercialization of microfluidic point-of-care diagnostic devices. *Lab Chip*, 12:2118–2134, 2012.

[CMK03] Sung Kwon Cho, Hyejin Moon, and Chang-Jin Kim. Creating, transporting, cutting, and merging liquid droplets by electrowetting-based actuation for digital microfluidic circuits. *IEEE Journal of Microelectromechanical Systems*, 12(1):70–80, 2003.

[CNFW12] Kihwan Choi, Alphonsus H. C. Ng, Ryan Fobel, and Aaron R. Wheeler. Digital microfluidics. *Annual Review of Analytical Chemistry*, 5(1):413–430, 2012.

[COW$^+$08] Fabienne Courtois, Luis F. Olguin, Graeme Whyte, Daniel Bratton, Wilhelm T. S. Huck, Chris Abell, and Florian Hollfelder. An integrated device for monitoring time-dependent in vitro expression from single genes in picolitre droplets. *ChemBioChem*, 9(3):439–446, 2008.

[cpl] IBM ILOG CPLEX Optimizer. http://www.ibm.com/software/integration/optimization/cplex/.

[CS07] Krishnendu Chakrabarty and Fei Su. *Digital Microfluidic Biochips - Synthesis, Testing, and Reconfiguration Techniques*. CRC Press, 2007.

[CTF$^+$02] Nick Christodoulides, Maiyen Tran, Pierre N. Floriano, Marc Rodriguez, Adrian Goodey, Mehnaaz Ali, Dean Neikirk, , and John T. McDevitt. A microchip-based multianalyte assay system for the assessment of cardiac risk. *Analytical Chemistry*, 74(13):3030–3036, 2002.

[CUQ01] Hou-Pu Chou, Marc A. Unger, and Stephen R. Quake. A microfabricated rotary pump. *Biomedical Microdevices*, 3(4):323–330, 2001.

[CWJ12] Chang-Yu Chen, Andrew M. Wo, and De-Shien Jong. A microfluidic concentration generator for dose-response assays on ion channel pharmacology. *Lab Chip*, 12:794–801, 2012.

[DCJW01] Stephan K. W. Dertinger, Daniel T. Chiu, Noo Li Jeon, and George M. Whitesides. Generation of gradients having complex shapes using microfluidic networks. *Analytical Chemistry*, 73(6):1240–1246, 2001.

[DM06] Petra S. Dittrich and Andreas Manz. Lab-on-a-chip: microfluidics in drug discovery. *Nature Reviews Drug Discovery*, 5:210–218, 2006.

[dMB08] Leonardo Mendonça de Moura and Nikolaj Bjørner. Z3: An efficient SMT solver. In *Proc. of TACAS*, 337–340, [Z3 is available at https://github.com/Z3Prover/z3], 2008.

[dMB11] Leonardo Mendonça de Moura and Nikolaj Bjørner. Satisfiability modulo theories: introduction and applications. *Commun. ACM*, 54(9):69–77, 2011.

[DVGL94] Gary V. Doern, Raymond Vautour, Michael Gaudet, and Bruce Levy. Clinical impact of rapid in vitro susceptibility testing and bacterial identification. *Journal of Clinical Microbiology*, 32(7):1757–1762, July 1994.

[DY11] Sabo Wada Dutse and Nor Azah Yusof. Microfluidics-based lab-on-chip systems in DNA-based biosensing: An overview. *Lab Chip*, 11:5754–5768, 2011.

[DYHH13] Trung Anh Dinh, Shigeru Yamashita, Tsung-Yi Ho, and Yuko Hara-Azumi. A clique-based approach to find binding and scheduling result in flow-based microfluidic biochips. In *Proc. ASP-DAC*, 199–204, 2013.

[DZN04] J. Ducrée, R. Zengerle, and J. D. Newman. *FlowMap: Microfluidics Roadmap for the Life Sciences*. FlowMap Consortium, 2004.

[EOJ+14] David Erickson, Dakota O'Dell, Li Jiang, Vlad Oncescu, Abdurrahman Gumus, Seoho Lee, Matthew Mancuso, and Saurabh Mehta. Smartphone technology can be transformative to the deployment of lab-on-chip diagnostics. *Lab Chip*, 14:3159–3164, 2014.

[Fai07] Richard B. Fair. Digital microfluidics: is a true lab-on-a-chip possible? *Microfluidics and Nanofluidics*, 3(3):245–281, 2007.

[FM11] Luis M. Fidalgo and Sebastian J. Maerkl. A software-programmable microfluidic device for automated biology. *Lab Chip*, 11:1612–1619, 2011.

[FPM12] D. Friedrich, C. P. Please, and T. Melvin. Design of novel microfluidic concentration gradient generators suitable for linear and exponential concentration ranges. *Chemical Engineering Journal*, 193–194:296–303, 2012.

[Fre61] Herbert Freeman. On the encoding of arbitrary geometric configurations. *IRE Transactions on Electronic Computers*, 10(2):260–268, 1961.

[GB14] Daniel Grissom and Philip Brisk. Fast online synthesis of digital microfluidic biochips. *IEEE Trans. on CAD*, 33(3):356–369, 2014.

[GCW+15] Daniel Grissom, Christopher Curtis, Skyler Windh, Calvin Phung, Navin Kumar, Zachary Zimmerman, Kenneth O'Neal, Jeffrey McDaniel, Nick Liao, and Philip Brisk. An open-source compiler and PCB synthesis tool for digital microfluidic biochips. *Integration*, 51:169–193, 2015.

[GHW+18] Anle Ge, Liang Hu, Xixian Wang, Jinchi Zhu, Xiaojun Feng, Wei Du, and Bi-Feng Liu. Logarithmic bacterial gradient chip for analyzing the effects of dietary restriction on C. elegans growth. *Sensors and Actuators B: Chemical*, 255(1):735–744, 2018.

[GSO+15] S. Garcia, R. Sunyer, A. Olivares, J. Noailly, J. Atencia, and X. Trepat. Generation of stable orthogonal gradients of chemical concentration and substrate stiffness in a microfluidic device. *Lab Chip*, 15:2606–2614, 2015.

[GV04] Peter R. C., Gascoyne, and Jody V. Vykoukal. Dielectrophoresis-based sample handling in general-purpose programmable diagnostic instruments. *Proc IEEE Inst. Electr. Electron. Eng.*, 92:22–42, 2004.

[GWY+17] A. Grimmer, Q. Wang, H. Yao, T.-Y. Ho, and R. Wille. Close-to-optimal placement and routing for continuous-flow microfluidic biochips. In *Proc. of ASP-DAC*, 530–535, 2017.

[HHC14] Yi-Ling Hsieh, Tsung-Yi Ho, and Krishnendu Chakrabarty. Biochip synthesis and dynamic error recovery for sample preparation using digital microfluidics. *IEEE Trans. on CAD*, 33(2):183–196, 2014.

[HLC12] Juinn-Dar Huang, Chia-Hung Liu, and Ting-Wei Chiang. Reactant minimization during sample preparation on digital microfluidic biochips using skewed mixing trees. In *Proc. ICCAD*, 377–384, 2012.

[HLH15] Chi-Mei Huang, Chia-Hung Liu, and Juinn-Dar Huang. Volume-oriented sample preparation for reactant minimization on flow-based microfluidic biochips with multi-segment mixers. In *Proc. of DATE*, 1114–1119, 2015.

[HLL13] Juinn-Dar Huang, Chia-Hung Liu, and Huei-Shan Lin. Reactant and waste minimization in multitarget sample preparation on digital microfluidic biochips. *IEEE Trans. on CAD*, 32(10):1484–1494, 2013.

[Ho14] Tsung-Yi Ho. Design automation for digital microfluidic biochips. *IPSJ Trans. System LSI Design Methodology*, 7:16–26, 2014.

[HQ03] Jong Wook Hong and Stephen R. Quake. Integrated nanoliter systems. *Nature Biotechnology*, 21:1179 –1183, 2003.

[HSH+04] Jong Wook Hong, Vincent Studer, Giao Hang, W French Anderson, and Stephen R Quake. A nanoliter-scale nucleic acid processor with parallel architecture. *Nature Biotechnology*, 22:435–439, 2004.

[HZC10] Tsung-Yi Ho, Jun Zeng Zeng, and Krishnendu Chakrabarty. Digital microfluidic biochips: A vision for functional diversity and more than *moore*. In *Proc. of ICCAD*, 578–585, Nov 2010.

[ILC15] Mohamed Ibrahim, Zipeng Li, and Krishnendu Chakrabarty. *Advances in Design Automation Techniques for Digital-Microfluidic Biochips*, 190–223. Springer Fachmedien Wiesbaden, 2015.

[InV] World in vitro diagnostics (IVD) market - opportunities and forecasts, 2013–2020. https://www.alliedmarketresearch.com/ivd-in-vitro-diagnostics-market.

[ITR] The international technology roadmap for semiconductors, 2007. http://www.itrs.net.

[JHK+11] Yun-Ho Jang, Matthew J. Hancock, Sang Bok Kim, Seila Selimovic, Woo Young Sim, Hojae Bae, and Ali Khademhosseini. An integrated microfluidic device for two-dimensional combinatorial dilution. *Lab Chip*, 11:3277–3286, 2011.

[Jov] Polymerase chain reaction: Basic protocol plus troubleshooting and optimization strategies. http://www.jove.com/video/3998/polymerase-chain-reaction-basic-protocol-plus-troubleshooting.

[KF08] Thomas M. Keenan and Albert Folch. Biomolecular gradients in cell culture systems. *Lab Chip*, 8:34–57, 2008.

[KLK+08] Choong Kim, Kangsun Lee, Jong Hyun Kim, Kyeong Sik Shin, Kyu-Jung Lee, Tae Song Kim, and Ji Yoon Kang. A serial dilution microfluidic device using a ladder network generating logarithmic or linear concentrations. *Lab-on-a-Chip*, 8(3):473–479, 2008.

[KLP+12] Sung Jin Kim, David Lai, Joong Y. Park, Ryuji Yokokawa, and Shuichi Takayama. Microfluidic automation using elastomeric valves and droplets: reducing reliance on external controllers. *Small (Weinheim an der Bergstrasse, Germany)*, 8:2925–2934, 2012.

[KRC+13] Srijan Kumar, Sudip Roy, Partha Pratim Chakrabarti, Bhargab B. Bhattacharya, and Krishnendu Chakrabarty. Efficient mixture preparation on digital microfluidic biochips. In *Proc. DDECS*, 205–210, 2013.

[KS08] Daniel Kroening and Ofer Strichman. *Decision Procedures - An Algorithmic Point of View*. Springer, 2008.

[KWD14] Oliver Keszocze, Robert Wille, and Rolf Drechsler. Exact routing for digital microfluidic biochips with temporary blockages. In *Proc. of ICCAD*, 405–410, 2014.

[KWHD14] Oliver Keszocze, Robert Wille, Tsung-Yi Ho, and Rolf Drechsler. Exact one-pass synthesis of digital microfluidic biochips. In *Proc. of DAC*, 1–6, 2014.

[LCG+13] Ming-Hua Lin, John Gunnar Carlsson, Dongdong Ge, Jianming Shi, and Jung-Fa Tsai. A review of piecewise linearization methods. *Mathematical Problems in Engineering*, 2013.

[LCg16] Yung-Chun Lei, Yi-Ling Chen, and Juinn-Dar Huang g. Reactant cost minimization through target concentration selection on microfluidic biochips. In *Proc. BioCAS*, 58–61, 2016.

[LCH15] Chia-Hung Liu, Ting-Wei Chiang, and Juinn-Dar Huang. Reactant minimization in sample preparation on digital microfluidic biochips. *IEEE Trans. on CAD*, 34(9):1429–1440, 2015.

[LCLH13] Chia-Hung Liu, Hao-Han Chang, Tung-Che Liang, and Juinn-Dar Huang. Sample preparation for many-reactant bioassay on dmfbs using common dilution operation sharing. In *Proc. ICCAD*, 615–621, 2013.

[LCWF11a] Chia-Yen Lee, Chin-Lung Chang, Yao-Nan Wang, and Lung-Ming Fu. Microfluidic mixing: A review. *International Journal of Molecular Sciences*, 12(5):3263–3287, 2011.

[LCWF11b] Chia-Yen Lee, Chin-Lung Chang, Yao-Nan Wang, and Lung-Ming Fu. Microfluidic mixing: A review. *Int. J. Mol. Sci.*, 12(5): 3263–3287, 2011.

[LDT] Architecture layout and dynamics of linear dilution gradient generation algorithm. http://www.iscal.ac.in/~sukanta_r/LinearDilutionGradient.zip.

[LKA+09] Kangsun Lee, Choong Kim, Byungwook Ahn, Rajagopal Panchapakesan, Anthony R. Full, Ledum Nordee, Ji Yoon Kang, and Kwang W. Oh. Generalized serial dilution module for monotonic and arbitrary microfluidic gradient generators. *Lab Chip*, 9:709–717, 2009.

[LKK+11] Jong Min Lee, Ji-eun Kim, Edward Kang, Sang-Hoon Lee, and Bong Geun Chung. An integrated microfluidic culture device to regulate endothelial cell differentiation from embryonic stem cells. *Electrophoresis*, 32(22):3133–3137, 2011.

[LLH16] Yung-Chun Lei, Tung-Hsuan Lin, and Juinn-Dar Huang. Multi-objective sample preparation algorithm for microfluidic biochips supporting various mixing models. In *Proc. of SOCC*, 96–101, 2016.

[LSH15] Chia-Hung Liu, Kuo-Cheng Shen, and Juinn-Dar Huang. Reactant minimization for sample preparation on microfluidic biochips with various mixing models. *IEEE Trans. on CAD*, 34(12):1918–1927, 2015.

[LYDP17] Zhao-Miao LIU, Yang YANG, Yu DU, and Yan PANG. Advances in droplet-based microfluidic technology and its applications. *Chinese Journal of Analytical Chemistry*, 45(2):282–296, 2017.

[MB05] Frieder Mugele and Jean-Christophe Baret. Electrowetting: from basics to applications. *Journal of Physics: Condensed Matter*, 17(28):R705–R774, 2005.

[MGC+11] Simone Luigi Marasso, Eros Giuri, Giancarlo Canavese, Riccardo Castagna, Marzia Quaglio, Ivan Ferrante, Denis Perrone, and Matteo Cocuzza. A multilevel lab on chip platform for DNA analysis. *Biomedical Microdevices*, 13(1):19–27, 2011.

[MHR+10] Daniel Mark, Stefan Haeberle, Günter Roth, Felix Von Stetten, and Roland Zengerle. Microfluidic lab-on-a-chip platforms: Requirements, characteristics and applications. *Chem. Soc. Rev.*, 39:1153–1182, 2010.

[MMVKI10] Andrew J. Muinonen-Martin, Douwe M. Veltman, Gabriela Kalna, and Robert H. Insall. An improved chamber for direct visualisation of chemotaxis. *PLoS ONE*, 5(12):1–9, 12 2010.

[MPP11] Wajid H. Minhass and Jan Madsen Paul Pop. System-level modeling and synthesis of flow-based microfluidic biochips. In *Proc. CASES*, 225–234, 2011.

[MQ07] Jessica Melin and Stephen R. Quake. Microfluidic large-scale integration: The evolution of design rules for biological automation. *Annual Review of Biophysics and Biomolecular Structure*, 36(1): 213–231, 2007.

[MRB+14] Debasis Mitra, Sudip Roy, Sukanta Bhattacharjee, Krishnendu Chakrabarty, and Bhargab B. Bhattacharya. On-chip sample preparation for multiple targets using digital microfluidics. *IEEE Trans. on CAD*, 33(8):1131–1144, 2014.

[MS09] Carol A. Meyers and Andreas S. Schulz. Integer equal flows. *Operations Research Letters*, 37(4):245–249, July 2009.

[NGL+12] Pavel Neuži, Stefan Giselbrecht, Kerstin Länge, Tony Jun Huang, and Andreas Manz. Revisiting lab-on-a-chip technology for drug discovery. *Nature Reviews Drug Discovery*, 11:620–632, 2012.

[NOT06] Robert Nieuwenhuis, Albert Oliveras, and Cesare Tinelli. Solving SAT and SAT Modulo Theories: From an abstract Davis–Putnam–Logemann–Loveland procedure to DPLL(T). *J. ACM*, 53(6):937–977, 2006.

[NTFF04] Christopher Neils, Zachary Tyree, Bruce Finlayson, and Albert Folch. Combinatorial mixing of microfluidic streams. *Lab-on-a-Chip*, 4(4):342–350, 2004.

[OMRW06] Adrian T. O'Neill, Nancy Monteiro-Riviere, and Glenn M. Walker. A serial dilution microfluidic device for cytotoxicity assays. In *Proc. EMBS*, 2836–2839, August 2006.

[PAC15] Paul Pop, Ismail Emre Araci, and Krishnendu Chakrabarty. Continuous-flow biochips: Technology, physical-design methods, and testing. *IEEE Design & Test*, 32(6):8–19, 2015.

[PCRa] PCR master mix calculator. http://www.mutationdiscovery.com.

[PCRb] PCR with Taq DNA Polymerase. http://www.protocols.io/view/PCR-with-Taq-DNA-Polymerase-M0273-imst9m.

[PFS00] Michael G. Pollack, Richard B. Fair, and Alexander D. Shenderov. Electrowetting-based actuation of liquid droplets for microfluidic applications. *Applied Physics Letters*, 77(11):1725–1726, 2000.

[RBC10] Sudip Roy, Bhargab B. Bhattacharya, and Krishnendu Chakrabarty. Optimization of dilution and mixing of biochemical samples using digital microfluidic biochips. *IEEE Trans. on CAD*, 29:1696–1708, Nov 2010.

[RBC11] Sudip Roy, Bhargab B. Bhattacharya, and Krishnendu Chakrabarty. Waste-aware dilution and mixing of biochemical samples with digital microfluidic biochips. In *Proc. DATE*, 1059–1064, 2011.

[RBGC14a] Sudip Roy, Bhargab B. Bhattacharya, Sarmishtha Ghoshal, and Krishnendu Chakrabarty. High-throughput dilution engine for sample preparation on digital microfluidic biochips. *IET Computers & Digital Techniques*, 8(4):163–171, 2014.

[RBGC14b] Sudip Roy, Bhargab B. Bhattacharya, Sarmishtha Ghoshal, and Krishnendu Chakrabarty. Theory and analysis of generalized mixing and dilution of biochemical fluids using digital microfluidic biochips. *ACM JETC*, 11(1):2:1–2:33, 2014.

[RCK+15a] Sudip Roy, Partha P. Chakrabarti, Srijan Kumar, Krishnendu Chakrabarty, and Bhargab B. Bhattacharya. Layout-aware mixture preparation of biochemical fluids on application-specific digital microfluidic biochips. *ACM TODAES*, 20(3):45.1–45.34, 2015.

[RCK+15b] Sudip Roy, Partha Pratim Chakrabarti, Srijan Kumar, Krishnendu Chakrabarty, and Bhargab B. Bhattacharya. Layout-aware mixture preparation of biochemical fluids on application-specific digital microfluidic biochips. *ACM Trans. Design Autom. Electr. Syst.*, 20(3):45, 2015.

[RIAM02] Darwin R. Reyes, Dimitri Iossifidis, Pierre-Alain Auroux, and Andreas Manz. Micro total analysis systems. 1. introduction, theory, and technology. *Analytical Chemistry*, 74(12):2623–2636, 2002.

[RKC+14] Sudip Roy, Srijan Kumar, Partha Pratim Chakrabarti, Bhargab B. Bhattacharya, and Krishnendu Chakrabarty. Demand-driven mixture preparation and droplet streaming using digital microfluidic biochips. In *Proc. DAC*, 144:1–144:6, 2014.

[Ros74] Azriel Rosenfeld. Digital straight line segments. *IEEE Trans. Computers*, 23(12):1264–1269, 1974.

[RRC+16] Shwathy Ramesan, Amgad R. Rezk, Kai Wei Cheng, Peggy P. Y. Chan, and Leslie Y. Yeo. Acoustically-driven thread-based tuneable gradient generators. *Lab Chip*, 16:2820–2828, 2016.

[RWX+09] Jing Ruan, Lihui Wang, Mingfei Xu, Daxiang Cui, Xiaomian Zhou, and Dayu Liu. Fabrication of a microfluidic chip containing dam, weirs and gradient generator for studying cellular response to chemical modulation. *Materials Science and Engineering: C*, 29(3):674–679, 2009.

[SBPH12] Ralf Seemann, Martin Brinkmann, Thomas Pfohl, and Stephan Herminghaus. Droplet based microfluidics. *Reports on Progress in Physics*, 75(1):016601:1–41, 2012.

[SBV11] Meng Sun, Swastika S. Bithi, and Siva A. Vanapalli. Microfluidic static droplet arrays with tuneable gradients in material composition. *Lab Chip*, 11:3949–3952, 2011.

[SFB14] Eric K. Sackmann, Anna L. Fulton, and David J. Beebe. The present and future role of microfluidics in biomedical research. *Nature*, 507:181–189, 2014.

[SHK10] Shinji Sugiura, Koji Hattori, and Toshiyuki Kanamori. Microfluidic serial dilution cell-based assay for analyzing drug dose response over a wide concentration range. *Analytical Chemistry*, 82(19):8278–8282, 2010.

[SHT+08] Ramakrishna Sista, Zhishan Hua, Prasanna Thwar, Arjun Sudarsan, Vijay Srinivasan, Allen Eckhardt, Michael Pollack, and Vamsee Pamula. Development of a digital microfluidic platform for point of care testing. *Lab Chip*, 8:2091–2104, 2008.

[SI03] Helen Song and Rustem F. Ismagilov. Millisecond kinetics on a microfluidic chip using nanoliters of reagents. *Journal of the American Chemical Society*, 125(47):14613–14619, 2003.

[SIP16a] Himali Somaweera, Akif Ibraguimov, and Dimitri Pappas. A review of chemical gradient systems for cell analysis. *Analytica Chimica Acta*, 907:7–17, 2016.

[SIP16b] Himali Somaweera, Akif Ibraguimov, and Dimitri Pappas. A review of chemical gradient systems for cell analysis. *Analytica Chimica Acta*, 907:7–17, 2016.

[SPF04] Vijay Srinivasan, Vamsee K Pamula, and Richard B Fair. Droplet-based microfluidic lab-on-a-chip for glucose detection. *Analytica Chimica Acta*, 507(1):145–150, 2004.

[SPO+08] Dean Y. Stevens, Camille R. Petri, Jennifer L. Osborn, Paolo Spicar-Mihalic, Katherine G. McKenzie, and Paul Yager. Enabling a microfluidic immunoassay for the developing world by integration of on-card dry reagent storage. *Lab Chip*, 8(12):2038–2045, 2008.

[SQH+15] Yi Sun, Than Linh Quyen, Tran Quang Hung, Wai Hoe Chin, Anders Wolffa, and Dang Duong Bang. A lab-on-a-chip system with integrated sample preparation and loop-mediated isothermal amplification for rapid and quantitative detection of *salmonella* spp. in food samples. *Lab Chip*, 15:1898–1904, 2015.

[SWJ08] Kang Sun, Zongxing Wang, and Xingyu Jiang. Modular microfluidics for gradient generation. *Lab Chip*, 8:1536–1543, 2008.

[SWLJ06] Wajeeh Saadi, Shur-Jen Wang, Francis Lin, and Noo Li Jeon. A parallel-gradient microfluidic chamber for quantitative analysis of breast cancer cell chemotaxis. *Biomedical Microdevices*, 8(2):109–118, 2006.

[Tar72] Robert E. Tarjan. Depth-first search and linear graph algorithms. *SIAM J. Comput.*, 1(2):146–160, 1972.

[TGZ+17] Jingxuan Tian, Yibo Gao, Bingpu Zhou, Wenbin Cao, Xiaoxiao Wu, and Weijia Wen. A valve-free 2D concentration gradient generator. *RSC Adv.*, 7:27833–27839, 2017.

[TLHS15] Tsun-Ming Tseng, Bing Li, Tsung-Yi Ho, and Ulf Schlichtmann. Reliability-aware synthesis for flow-based microfluidic biochips by dynamic-device mapping. In *Proc. DAC*, 141:1–141:6, 2015.

[TLL+16] Tsun-Ming Tseng, Mengchu Li, Bing Li, Tsung-Yi Ho, and Ulf Schlichtmann. Columba: co-layout synthesis for continuous-flow microfluidic biochips. In *Proc. DAC*, 147:1–147:6, 2016.

[TLSH15] Tsun-Ming Tseng, Bing Li, Ulf Schlichtmann, and Tsung-Yi Ho. Storage and caching: Synthesis of flow-based microfluidic biochips. *IEEE Design & Test*, 32(6):69–75, 2015.

[TUTA08] William Thies, John Paul Urbanski, Todd Thorsen, and Saman P. Amarasinghe. Abstraction layers for scalable microfluidic biocomputing. *Natural Computing*, 7(2):255–275, 2008.

[USP] General notices and requirements: Applying to standards, tests, assays, and other specifications of the united states pharmacopeia. http://www.usp.org/sites/default/files/usp_pdf/EN/USPNF/USP34-NF29GeneralNotices.pdf.

[UTR+06] John Paul Urbanski, William Thies, Christopher Rhodes, Saman Amarasinghe, and Todd Thorsen. Digital microfluidics using soft lithography. *Lab Chip*, 6:96–104, 2006.

[VCLBPT10] Guilhem Velve-Casquillas, Mael Le Berre, Matthieu Piel, and Phong Tran. Microfluidic tools for cell biological research. *Nano Today*, 5(1):28–47, 2010.

[WBG16] Junchao Wang, Philip Brisk, and William H. Grover. Random design of microfluidics. *Lab Chip*, 16:4212–4219, 2016.

[WC10] Donald Wlodkowic and Jonathan M. Cooper. Microfabricated analytical systems for integrated cancer cytomics. *Analytical and Bioanalytical Chemistry*, 398(1):193–209, 2010.

[WCX+15] Hao Wang, Chia-Hung Chen, Zhuolin Xiang, Ming Wang, and Chengkuo Lee. A convection-driven long-range linear gradient generator with dynamic control. *Lab Chip*, 15:1445–1450, 2015.

[WJWL10] Shiying Wang, Ning Ji, Wei Wang, and Zhihong Li. Effects of non-ideal fabrication on the dilution performance of serially functioned microfluidic concentration gradient generator. In *Nano/Micro Engineered and Molecular Systems (NEMS)*, 169–172, 2010.

[WLP17] Xiang Wang, Zhaomiao Liu, and Yan Pang. Concentration gradient generation methods based on microfluidic systems. *RSC Adv.*, 7: 29966–29984, 2017.

[WMRRN10] Glenn M. Walker, Nancy Monteiro-Riviere, Jillian Rouse, and O'Adrian T. Neill. A linear dilution microfluidic device for cytotoxicity assays. In *Lab Chip*, volume 7, 226–232, 2010.

[WSR+13] Judyta Wegrzyn, Adam Samborski, Louisa Reissig, Piotr M. Korczyk, Slawomir Blonski, and Piotr Garstecki. Microfluidic architectures for efficient generation of chemistry gradations in droplets. *Microfluidics and Nanofluidics*, 14(1):235–245, January 2013.

[XCP10] Tao Xu, Krishnendu Chakrabarty, and Vamsee K. Pamula. Defect-tolerant design and optimization of a digital microfluidic biochip for protein crystallization. *IEEE Trans. on CAD*, 29(4):552–565, 2010.

[YEF$^+$06] Paul Yager, Thayne Edwards, Elain Fu, Kristen Helton, Kjell Nelson, Milton R. Tam, and Bernhard H. Weigl. Microfluidic diagnostic technologies for global public health. *Nature*, 442:412–418, 2006.

[YHC15] Hailong Yao, Tsung-Yi Ho, and Yici Cai. PACOR: practical control-layer routing flow with length-matching constraint for flow-based microfluidic biochips. In *Proc. DAC*, 142:1–142:6, 2015.

[ZWQ12] Guoxia Zheng, Yunhua Wang, and Jianhua Qin. Microalgal motility measurement microfluidic chip for toxicity assessment of heavy metals. *Analytical and Bioanalytical Chemistry*, 404(10):3061–3069, 2012.

[15] Preparation of Plasmid DNA by Alkaline Lysis with SDS: Minipreparation, Cold Spring Harb Protocols. http://cshprotocols.cshlp.org/content/2006/1/pdb.prot4084.citation.

[UMK$^+$09] A. G. Uren, H. Mikkers, J. Kool, L. van der Weyden, A. H. Lund, C. H. Wilson, R. Rance, J. Jonkers, M. van Lohuizen, A. Berns, and D. J. Adams. A high throughput splinkerette-pcr method for the isolation and sequencing of retroviral insertion sites. *Nature Protocols*, 4(5):789–798, 2009.

[17] Openwetware, 2009. http://openwetware.org/wiki/Main_Page.

Index

About the Authors

Sukanta Bhattacharjee received the B.Tech. degree in computer science and engineering from the University of Calcutta, India and the M.Tech. and Ph.D. degrees in computer science from the Indian Statistical Institute, Kolkata, India. He is currently working as a post-doctoral associate in the Center for Cyber Security, New York University, Abu Dhabi.

His research interests include design automation algorithms for microfluidic biochip, formal methods, and security.

Bhargab B. Bhattacharya had been on the standing faculty of Computer Science and Engineering at the Indian Statistical Institute, Kolkata, during 1982–2018, where currently, he is associated as Honorary Visiting Professor after his retirement. He received the B.Sc. degree in physics, the B.Tech. and M.Tech. degrees in radiophysics and electronics, and the Ph.D. degree in computer science, all from the University of Calcutta.

He held visiting professorship at the University of Nebraska-Lincoln, and at Duke University, USA, at the University of Potsdam, Germany, at the Kyushu Institute of Technology, Japan, at Tsinghua University, Beijing, China, at IIT Kharagpur and IIT Guwahati, India. His current research interest includes design and test of integrated circuits and microfluidic biochips. He is a Fellow of the Indian National Academy of Engineering (INAE), a Fellow of the National Academy of Sciences (India), and a Fellow of the IEEE. He was named INAE Chair Professor (2016–18), and AICTE-INAE Distinguished Visiting Professor (2018–20).

Krishnendu Chakrabarty received the B.Tech. degree from the Indian Institute of Technology, Kharagpur, in 1990, and the M.S.E. and Ph.D. degrees from the University of Michigan, Ann Arbor, in 1992 and 1995, respectively. He is now the William H. Younger Distinguished Professor and Department Chair of Electrical and Computer Engineering and Professor of Computer Science at Duke University.

Prof. Chakrabarty is a recipient of the National Science Foundation CAREER award, the Office of Naval Research Young Investigator award, the Humboldt Research Award from the Alexander von Humboldt Foundation, Germany, the IEEE Transactions on CAD Donald O. Pederson Best Paper Award the ACM Transactions on Design Automation of Electronic Systems Best Paper Award and over a dozen best paper awards at major conferences. He is also a recipient of the IEEE Computer Society Technical Achievement Award the IEEE Circuits and Systems Society Charles A. Desoer Technical Achievement Award and the Distinguished Alumnus Award from the Indian Institute of Technology, Kharagpur. He is a Research Ambassador of the University of Bremen, Germany, and a Hans Fischer Senior Fellow at the Institute for Advanced Study, Technical University of Munich, Germany. He is also a recipient of the Japan Society for the Promotion of Science (JSPS) Prestigious Fellowship in the Short-Term S-Category. Prof. Chakrabartys current research interests include: testing and design-for-testability of integrated circuits and systems; digital microfluidics, biochips, and cyberphysical systems; data analytics for fault diagnosis, failure prediction, anomaly detection, and hardware security; neuromorphic computing systems. He is a Fellow of ACM and a Golden Core Member of the IEEE Computer Society.